Wafer Manufacturing

Wafer Manufacturing

Shaping of Single Crystal Silicon Wafers

Imin Kao
SUNY at Stony Brook University
USA

Chunhui Chung
National Cheng Kung University
Taiwan

This edition first published 2021
© 2021 John Wiley & Sons Ltd

The right of Imin Kao and Chunhui Chung to be identified as the authors of this work has been asserted in accordance with law.

Registered Offices
John Wiley & Sons, Inc., 111 River Street, Hoboken, NJ 07030, USA
John Wiley & Sons Ltd, The Atrium, Southern Gate, Chichester, West Sussex, PO19 8SQ, UK

Editorial Office
The Atrium, Southern Gate, Chichester, West Sussex, PO19 8SQ, UK

For details of our global editorial offices, customer services, and more information about Wiley products visit us at www.wiley.com.

Wiley also publishes its books in a variety of electronic formats and by print-on-demand. Some content that appears in standard print versions of this book may not be available in other formats.

Library of Congress Cataloging-in-Publication Data

Names: Kao, Imin, author. | Chung, Chunhui, author.
Title: Wafer manufacturing : shaping of single crystal silicon wafers /
    Imin Kao, Chunhui Chung.
Description: First edition. | Hoboken, NJ : Wiley, 2021. | Includes
    bibliographical references and index.
Identifiers: LCCN 2020034639 (print) | LCCN 2020034640 (ebook) | ISBN
    9780470061213 (hardback) | ISBN 9781118696255 (adobe pdf) | ISBN
    9781118696231 (epub) | ISBN 9781118696224 (obook)
Subjects: LCSH: Semiconductor wafers–Design and construction.
Classification: LCC TK7871.85 .K295 2021 (print) | LCC TK7871.85 (ebook)
    | DDC 621.3815/2–dc23
LC record available at https://lccn.loc.gov/2020034639
LC ebook record available at https://lccn.loc.gov/2020034640

Cover Design: Wiley
Cover Image: © kynny / Getty Images

Set in 9.5/12.5pt STIXTwoText by SPi Global, Chennai, India
Printed and bound by CPI Group (UK) Ltd, Croydon, CR0 4YY

10 9 8 7 6 5 4 3 2 1

Contents

Preface

This book is designed and written with both research and engineering practice in mind. Subjects pertaining to the fundamentals of wafer manufacturing and recent research advances are presented, as well as practical engineering and technology for practitioners in this industry. This book is ideal for a novice who wants to gain insightful information and knowledge about wafer manufacturing, and to learn how the prime (or premium) wafers that are used in microelectronics fabrication are produced. At the same time, this book is full of research topics and presentations of advances in fundamental knowledge about relevant manufacturing processes such that it will meet the needs of curious minds who want to investigate issues beyond the status quo and to study these processes with an objective to improve a myriad of related manufacturing processes.

Wafer manufacturing has been a field in which experience and engineering know-how has played a significant role and was of instrumental value to the industry. With the emergence of new technology and research, the challenges are to understand fundamental process modeling in various manufacturing processes of wafer manufacturing, and to be able to apply this new knowledge in the process control and management of wafering processes. With the familiarity of both research issues and practical engineering, the authors strive to achieve the goals stated above.

The book is a must-have for engineers and researchers in the field of wafer production and manufacturing. Although single crystalline silicon wafers are the majority of wafering done today in the industry, this book is also equally applicable, in many regards, to the wafer production of various crystal ingots such as III–V and II–VI compounds, silicon carbide, lithium niobates, and others.

It has been a major undertaking to commit to writing a book like this, although I have been conducting research in wafer manufacturing since 1994. This project has taken more time than was estimated initially when the first agreement was signed with the Wiley publisher. My goal was and still is to make this book a resource for those who want to understand the many aspects of wafer manufacturing technologies, as well as for those practitioners who want to utilize the knowledge to improve the existing processes.

Strict quality control has been imposed on the content of this book and subjects have been added or removed along the process of writing to make it as relevant as possible, based on the best judgment in the field. This book is put together with the best intent and effort, although it is far from being perfect. I welcome feedback and comments from readers, who can email me at WaferMfgBookKao@gmail.com, or send postal mail to: Professor Imin Kao, Department of Mechanical Engineering, Stony Brook, NY 11794-2300, USA.

Acknowledgement

This book is a culmination of many years of research on the subject, and has had the help of many colleagues, students, and mentors along the path of my academic career. Dr. Tai-ran Hsu, who gave me an opportunity as an assistant professor at San Jose State University, is a mentor and an inspiration when he endeavored to conduct research and study in MEMS in 1991, with the outcome of a book in MEMS (McGraw-Hill 2002). My research at Stony Brook University (SBU, State University of New York, SUNY) on the topic of wafer manufacturing started in 1994 when I joined SBU. I started this research with a grant from the DoE and collaborated with Dr Vish Prasad on innovative growth techniques of photovoltaic wafers, and slicing using a slurry wiresaw, which was a new machine tool for slicing ingots to wafers at that time. The research continued to be funded by NSF and industrial grants. The collaboration with Dr Fu-pen Chiang, who is a reputed and renowned researcher in optical metrology, led to work in novel wafer surface measurements and real-time monitoring of wire wear and wire web using optical metrology. Many PhD and MS students of mine worked on pioneering modeling and research on slurry wiresaws and wafer manufacturing that led to some of the contents of this book. I am grateful to have the dedication and contributions of Drs Milind Bhagavat (who started the concept of this book with me), Liqun Zhu, Songbin Wei, Sumeet Bhagavat, Chunhui Chung (who is a coauthor of this book), and Liming Li.

In addition, I would like to thank Professor Chao-Chang Arthur Chen of the National Taiwan University of Science and Technology and the Global Wafer Co., Limited in Taiwan that provided images of wafer manufacturing processes for my book. The images are used to illustrate various wafer manufacturing processes, from crystal growth to polished prime wafers, and make for a better presentation and relevance of the manufacturing processes.

I was fortunate to be working with some of the best colleagues at SBU when I was asked to step into various roles in administration, and participated in the establishment of innovative education initiatives, such as the Undergraduate Colleges system, new departments, and SUNY Korea – a global campus of SBU at Incheon, Korea. My mentor, Dr Yacov Shamash, helped me to appreciate the rigours in academic administration. I took a six-month sabbatical leave in 2016 at SUNY Korea to carry out my role as the Executive Director of SUNY Korea Academic Programs, and to spend more time on writing this book. I am very appreciative of the then President ChoonHo Kim with whom I partnered and participated in the daunting work, with a broader team at SBU and in Korea, in establishing a global university in a new mold with an academia–government–industry trilateral collaboration at an international level. Juggling administrative tasks with my research has been an intriguing challenge for me over the years, nonetheless.

Most importantly, I am forever grateful to my wife and life-long partner, Elaine Chang, who has been a stalwart supporter for me over more than 37 years of marriage. Without her unwavering support and considerable influence, this book would not have been possible. We have three children, Drs Katherine Kao, Jonathan Kao, and Emmeline Kao, who have helped with reading this work and have commented on it. I am indeed blessed to have my family and God in my life.

February 2020
Imin Kao
at Stony Brook, New York

Part I

From Crystal to Prime Wafers

1

Wafers and Semiconductors

1.1 Introduction

The fundamental building block from which microelectronic integrated circuit (IC) chips, MEMS (micro electro mechanical systems) devices, and microelectronic devices and systems are constructed is called a "wafer." This book discusses the nuances of the wafer manufacturing (or wafer production) technology and process, as well as the recent research and development contributing to the growth of the technologies in wafer manufacturing. Each chapter has a References section at the end to provide a list of books and articles for reference and further reading or study on the subjects presented.

This chapter provides basic information about wafers and the semiconductor revolution that has influenced our modern daily lives, including the subjects of silicon, wafers, mechanical materials, and surface properties.

1.2 Semiconductor Revolution

1.2.1 Classification of Materials

Materials are classified, based on their electrical conductivity, into the following four categories: conductor, insulator, superconductor, and semiconductor. Table 1.1 lists a few selected materials in the three most common categories and their resistivity values in Ω m. The resistance of a material, R, is defined as follows

$$R = \rho \frac{l}{A} \tag{1.1}$$

where ρ is the resistivity and l and A are the length and cross-sectional area of the material, respectively. When a voltage is applied to a material, it may cause electrical current to flow, depending on the resistivity of the material. The electrical resistivity is thus defined as the ratio between the electric field and the density of the current it creates, as in the following equation

$$\rho = \frac{E}{J} \tag{1.2}$$

Wafer Manufacturing: Shaping of Single Crystal Silicon Wafers,
First Edition. Imin Kao and Chunhui Chung.
© 2021 John Wiley & Sons Ltd. Published 2021 by John Wiley & Sons Ltd.

where E is the electric field applied inside the material with a unit of V m^{-1}, and J is the current density in A m, also measured inside the material. The conductivity is the inverse of resistivity as follows

$$\sigma = \frac{1}{\rho}. \tag{1.3}$$

Four classifications of material are described in the following.

1. Conductor – these materials conduct current when a voltage is applied. Examples include metals, many metal alloys, and carbon. Metals are the best naturally occurring conductors of electricity, because of their metallic bonding, and thus have low resistivity (see Table 1.1).

2. Insulator – these materials conduct no current when a voltage is applied. Nevertheless, the current will suddenly pass through the material if the voltage is high enough, for example, in the form of an arc or plasma. An insulator is sometimes referred to as a dielectric, which describes non-conduction of a direct current. The dielectric strength of an insulating material, then, is the electrical potential required to break down the insulator per unit thickness, with typical unit in V m^{-1} (or V in^{-1}). Examples of insulator materials include silicon dioxide (SiO_2), glass, ceramic materials, polymers, etc. Most ceramics and polymers, whose electrons are tightly bound by covalent and/or ionic bonding, are typically poor conductors. Many of these materials are used as insulators because they possess high resistivity values (See Table 1.1).

3. Superconductor – a superconductor is a material that exhibits zero resistivity. Superconductivity is a phenomenon that is observed in certain materials at low temperatures near absolute zero at which zero electrical resistance and expulsion of magnetic fields take place in certain materials. Such materials hold great interest and promise for many applications such as power transmission, magnetic levitation, and electronic switching.

4. Semiconductor – a semiconductor is dielectric in its natural state, but can conduct electricity under some conditions, either due to the addition of an impurity or because of temperature effects, but not others. Such characteristics make semiconductor materials a good medium for the control of electrical current. The resistivity of a semiconductor lies between that of insulators and conductors. Typical resistivities of semiconductor materials range from 10 to 10^5 Ω m.

Table 1.1 Resistivity of selected materials in Ω m, as defined in Equation (1.1). Superconductors have zero resistivity.

Conductors	*Resistivity*	Insulators	*Resistivity*	Semiconductors	*Resistivity*
Aluminum	2.8×10^{-8}	Polyurethane	1.2×10^{14}	Silicon	1.0×10^3
Silver	1.6×10^{-8}	Rubber	10^{12}–10^{15}	Germanium	5×10^{-1}
Copper	1.7×10^{-8}	SiO_2	10^3–4×10^{11}	GaAs	10^{-6}–10^{-2}
Carbon	5×10^{-5}	Glass	10^{10}–2×10^{14}	SiC	10^0–10^4

Values of resistivity vary depending on condition and environment, such as impurity and temperature. The commonly used unit in semiconductors is Ω cm, with 1 Ω m = 100 Ω cm.

Semiconductors are the foundation of modern electronics, including computers, electronic devices, and smart phones. Semiconductor solar photovoltaic panels directly convert light energy into electricity.

Common semiconducting materials are crystalline solids. Silicon is the most popular commercial semiconductor material. Many other materials are used, including germanium, gallium arsenide (GaAs), indium phosphide (InP), and silicon carbide (SiC). The electronic properties and the conductivity of a semiconductor can be changed by "doping," referring to a diffusion process that adds other materials to the semiconductor materials. This is typically achieved during crystalline silicon growth by adding boron or phosphorus impurities to the melt and then allowing it to solidify into silicon during the crystal growth process.

1.2.2 Semiconductor Revolution Today

Presently we are in the midst of a "semiconductor revolution." Many objects with which we are associated today have either semiconductor devices built in or have been manufactured with a machine that contains in-built semiconductor devices. These modern times are the "semiconductor era", akin to the "steel era" that propelled the "second industrial revolution" in the 19th century. In the last several decades, semiconductor technology has grow in leaps and bounds, partly driven by consumerism generated by telecommunication, wireless internet, and information technology and partly by the need for automation catering to mass or batch production and manufacturing of technologically sophisticated products. Thanks to semiconductor technology, we are now able to manufacture very complex products in large quantities at lower costs.

Today's world is embraced by three technologies, each in a different stage of maturity:

1. The microelectronics technology with integrated circuits (ICs) has been at the forefront of all the breath-taking achievements realized in the past few decades. At the core of IC technology is the development of very tiny but extremely robust electrical circuits that can act as the brain of a machine, such as that in a controller, computer, or smart devices. The road of microelectronics technology was paved with the invention of the first transistor in 1947 [American Physical Society (2000); Riordan and Hoddeson (1997)], as shown in Figure 1.1. The batch fabrication of transistors using planar technology enables the IC technology we know today. IC technology has been growing rapidly since the 1970s and since that time it has never shown any sign of decline. Radically new technologies and new products are still being introduced into the market every year and new concepts and innovations are being researched that will keep this technology moving for years to come.

2. The fabrication technology of IC has given birth to the world of MEMS (micro electro mechanical systems) which integrates the electrical power of ICs with micron-scale mechanical structures to build powerful applications pertaining to the real world. An example is air-bag sensors in automobiles. MEMS technology is in an exciting phase; it has already been utilized in everyday applications

Figure 1.1 The first transistor, made of germanium, at the AT&T Bell Laboratory. (Source: AT&T)

and is still concurrently looking for new applications. The "satellite navigation" industry and the "computer gaming industry" have already presented some new avenues, and there are undoubtedly many more opportunities to realize.

3. On the horizon is the new enticing world of "nanotechnology", which encompasses both fundamental research and promising applications in fields such as medicine, energy, biomedical, etc. At the root of nanotechnology is the confluence of electrical, mechanical, chemical, physical, and biological sciences and engineering. The fields of nanotechnology hold a lot of future promise.

The most fundamental part of the semiconductor world is the "semiconductor material" used to realize the technologies. In layman's parlance, semiconductors can show electrical conductivity between conductors and insulators, as described in Section 1.2.1. This property by itself is not enough to realize the technologies (ICs and MEMS) we know of today. What really drives today's semiconductor world are the properties of some of the semiconductors that allow their conductivity to be manipulated by either electrical signals or by processing. This is the property that enables the working of "transistors" which are the fundamental building blocks of semiconductor devices. Some semiconductor materials that have found their way into device applications are silicon, germanium, silicon carbide, gallium arsenide, indium phosphide, lithium niobate, and many more. By far, silicon is the most popular semiconductor material. Silicon meets more than 95% of the world's semiconductor needs.

1.2.3 Silicon Wafers and Solar Cells

Silicon as a semiconductor material has also found application in solar cells. As the world is facing the prospect of depletion of conventional energy sources, more emphasis is being put on harnessing unconventional energy such as solar energy. Silicon is the backbone of "solar cell" technology. Many of the demands that are put on silicon wafers in semiconductors are also applicable to the solar world. The only place where the solar world diverges from semiconductors is that semiconductor applications need single-crystalline (or mono-crystal) materials, while the solar cells can use single- or quasi-single- or poly-crystalline materials. Still, there is a lot of similarity between the manufacturing of wafers for IC chips and for solar cells. The mechanical properties needed for a semiconductor wafer or solar wafer are more or less similar. This book also discusses the wafer manufacturing issues associated with the solar industry.

1.3 Silicon Wafers Used in Device Manufacturing (IC and MEMS)

Silicon is a blue-gray, brittle chemical element in the same group as carbon and germanium in the periodic table and shares similar properties to microelectronic materials. Figure 1.2 shows pieces of high-purity, ready to melt silicon for growing boules. Comprising 28% of the earth's elements, silicon is second only to oxygen as the most common element on the surface of the earth. Silicon is a key ingredient in glasses, quartz, soil, and sand. As a brittle material, silicon has hardness and structural strength comparable to many metals such as aluminum and bronze. Semiconductor devices, or chips, are fabricated on a silicon substrate in the form of a wafer. The manufacture of silicon wafers begins with the growth of a single-crystalline silicon ingot (or boule). A single-crystalline silicon crystal is a boule composed of atoms arranged in a three-dimensional periodic pattern that extends throughout the

Figure 1.2 Silicon with purity higher than 99.9%. Source: Imin Kao.

material, required for semiconductor fabrication. Polycrystalline silicon crystals are formed with random orientations, and are therefore cheaper to manufacture. They are often used in photovoltaic solar devices.

1.3.1 Standard Wafer Diameters and Sizes

Silicon wafers are available in a variety of diametral sizes from 25.4 mm (1 in) to 300 mm (12 in) [Gise and Blanchard (1986)], and the future wafer size of 450 mm (18 in). Semiconductor fabrication plants (also known as fabs) are defined by the size of wafers that they are tooled to produce. Silicon wafers have historically consisted of diameters/sizes shown in Table 1.2. The size has gradually increased to improve throughput and reduce cost with current state-of-the-art semiconductor fabrication plants (fabs) processing wafers with a diameter of 300 mm (or 12 in). The next standard wafer size, as promoted by Intel, Samsung, TSMC, and SEMATECH, will be 450 mm (18 in) [SEMI (2012a); Young (2011a); Capraro (2013); Intel News Release (2008); Watanabe and Kramer (2006)]. This "industry transition to 450 mm wafers for leading-edge chip manufacturing will be one of the most complex and costly decisions in the history of semiconductors" [SEMI (2012b)]. According to SEMAT-ECH and SEMI, the "450 Consortia," a consortium of researchers in semiconductor technology, has been effective at creating a level platform for consensus building to leverage experience and expertise of a larger community, including program scopes in equipment, materials, guidelines, and testing [Young (2011b)]. In recent years, the Global 450 Consortium (G450C) has been established as a 450 mm wafer and equipment development program that is leveraging industry and government investments to demonstrate 450 mm process capabilities at the Albany NanoTech Complex

Table 1.2 Wafer diameters and thickness.

Wafer diameter in SI units	Equivalent wafer size	Typical wafer thickness
20 mm	1 inch	–
51 mm	2 inch	275 μm
76 mm	3 inch	375 μm
100 mm	4 inch	525 μm
125 mm	4.9 inch	625 μm
130 mm	5 inch	625 μm
150 mm	6 inch	675 μm
200 mm	8 inch	725 μm
300 mm	12 inch	775 μm
450 mm	18 inch	925 μm

The largest current standard size of wafers is 300 mm, with the next size of 450 mm expected in the near future. The unit of μm represents micrometer, or micron, with $1 \, \mu m = 0.001 \, mm = 1 \times 10^{-6} \, m$.

of the Colleges of Nanoscale Science and Engineering (CNSE) of SUNY Polytechnic Institute (SUNY Poly) [CNSE (2012); CNSE (2015)].

The 450 mm wafer is not simply an extension of the current technology for the 300 mm wafer. Reducing the defects of a 450 mm wafer is just as critical as discovering numerous technological breakthroughs and innovations.

Four factors are often mentioned in regard to improving the productivity of semiconductor manufacturing. They are

(i) Increasing capital equipment utilization
(ii) Improving die yield
(iii) Enlarging wafer sizes
(iv) Shrinking device feature sizes.

While the device feature sizes continue to decrease, a SEMATECH study by Anderson (1997) indicates that the size of wafers in microelectronics fabrication has remained unchanged for approximately 24 years, allowing a return on investment (ROI) of R&D and capital investment of the equipment and technology. Nevertheless, the semiconductor industry needs to increase wafer sizes because of the overall cost reduction resulting from a larger percentage of usable real estate of wafer surface and more dice[1] per wafer. The overall cost of manufacturing is reduced by using a larger wafer surface because more dice are produced using the same number of fabrication process steps. In addition, wafer size increases are also viewed in terms of the increase in wafer area. For example, the total surface area of a 300 mm (12 inch) wafer is 2.25 times larger than that of a 200 mm (8 inch) wafer, with an increase of 125% surface area. Likewise, the same increase of the wafer surface area is realized from 300 mm to 450 mm wafers.

1.3.2 Crystalline Orientation of Silicon Wafers

Table 1.3 illustrates the convention of wafer flats to indicate the crystallographic planes and doping for wafers with diameters 150 mm and smaller. The primary flat has the longest straight length along the circumference of the wafer, as shown at the bottom flats in the figure. The horizontal flats at the bottom of each wafer illustration indicate the major direction [1 1 0] along the primary flats. The secondary flat, if it exists, is the shorter straight length along the circumference of the wafer.

Modern wafers with a diameter of 200 mm or more use a notch to represent the crystallographic information with no visual indication of the doping type. Such a wafer with a notch is illustrated in Figure 1.3. The notch indicates the major direction [1 1 0] of the wafer, similar to the direction indicated by the primary flat in smaller wafers.

Wafers made of materials other than silicon may have different sizes and thicknesses as compared to the electronic silicon wafer. Wafer thickness is determined by the mechanical properties of the material, especially in bending with loading

---

1 When a wafer has been diced, the pieces left on the dicing tape are referred to as *die, dice* or *dies*.

Table 1.3 Convention of the orientation flats for wafers with diameters of 150 mm or smaller. The longer side at the bottom is the "primary" flat; the shorter side is the "secondary" flat.

Wafer	p-type	n-type
(1 0 0)	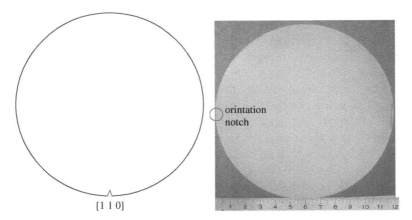 90° [1 1̄ 0]	180° [1 1̄ 0]
(1 1 1)	[1 1̄ 0]	45° [1 1̄ 0]

The Miller index is used here to denote plane with (…), and direction with […].

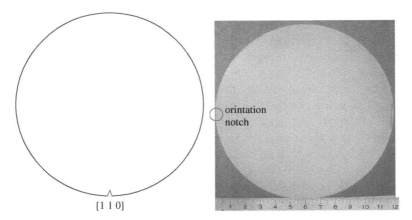

[1 1 0]

orintation notch

Figure 1.3 (Left) Convention of the orientation notch, indicating the primary orientation of [1 1 0] for wafers with a diameter of 200 mm or larger. (Right) A photo of an as-sliced silicon wafer of diameter 300 mm (12 in), with the orientation notch at the left side of the wafer. Source: Imin Kao.

and deflection due to its own weight, so as not to crack or break during processing and handling in microelectronics fabrication. More details will be discussed in Section 1.5.3.

1.3.3 Moore's Law

Moore's Law (as it is commonly referred to) is an empirical observation and prediction of the scale and the number of transistors on chips or integrated circuits, which roughly doubles every two years [Moore (1965)]. This prediction has proven to be accurate for nearly five decades, partly because it has been used in the semiconductor industry to guide long-term planning and to establish milestones for research and development. For example, an Intel website recently stated that "for decades, Intel has met this formidable challenge through investments in technology and manufacturing resulting in the unparalleled silicon expertise that has made Moore's Law a reality." [Intel Corp. (2005)]. A plot of CPU microprocessor transistor counts from 1971 to 2018 is shown in Figure 1.4. Note the vertical axis is in logarithmic scale. The line in Figure 1.4 corresponds to a growth that doubles every two years, as predicted by Moore's law.

Although Moore's law was initially a prediction of the growth of the scale of transistors in microprocessors, this principle has been employed in various technologies and industries, typically as a measure in digital technology for the rate of growth in size, density, speed, capacity, etc., based on this binary power law.

1.4 Surface Properties and Quality Measurements of Wafers

The processes and quality of wafer manufacturing (especially semiconductor wafers) have been improved over the last decades in response to the innovative pace and rigorous requirements of the surface finish of semiconductor and solar industries. Several important properties of wafer surface properties and quality measurements will be discussed in the following sections.

Conventionally, wafer surfaces are measured and characterized by the parameters of total thickness variation (TTV), warp, bow, flatness, and waviness. Typically, the unit of TTV, warp, and bow is microns. As the wafer diameters become larger, the requirements of surface finish become more stringent and it remains important to maintain high yield in electronic fabrication. A few important surface properties are discussed in the following subsections.

1.4.1 Surface Waviness: TTV, Bow, and Warp

The bulk properties that characterize the surface of wafers are best illustrated by Figure 1.5 in which a wafer spins between two capacitive probes that sample points on the wafer surfaces. Different surface measures are defined as presented in the following, based on various ASTM standards: for TTV in [ASTM Standard

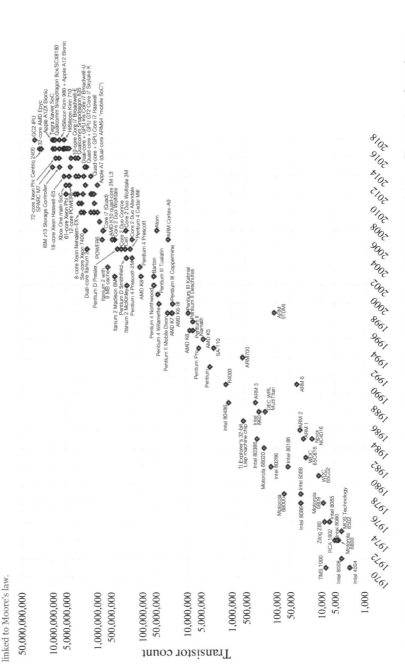

Figure 1.4 Plot of CPU microprocessor transistor counts from 1971 to 2018 and Moore's Law. Source: Moore's Law – The number of transistor on integrated circuit chips (1971–2018). Retrieved from: https://ourworldindata.org/uploads/2019/05/Transistor-Count-over-time-to-2018.png. Licensed under CCBY SA.

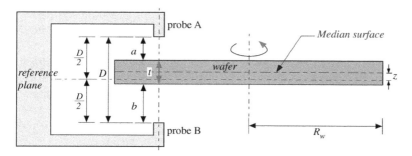

Figure 1.5 Illustration of a wafer under non-contact characterization of surface measurement. The probes A and B are typically a pair of capacitive probes that detect the distances a and b in automated non-contact scanning.

F1390 (2002); ASTM Standard F533 (2002); ASTM Standard F657 (1999)], for bow in [ASTM Standard F534 (2002)], and for warp in [ASTM Standard F1390 (2002); ASTM Standard F657 (1999)]. These ASTM standards had been withdrawn in May 2003, and were transferred to SEMI (www.semi.org). The SEMI documents describing the TTV, bow and warp can be found in [SEMI 3D12 (2015); SEMI MF1390 (2018); SEMI MF1530 (2018); SEMI MF533 (2010); SEMI MF534 (2014); SEMI MF657 (2014)].

The terminologies of TTV, bow, and warp, used to define wafer surface properties, are often referred to when describing the quality of surface finish of wafers. The following terms are first defined for the description of the various surfaces of a wafer.

- *Front surface* – the exposed surface upon which active semiconductor devices have been or will be fabricated.
- *Back surface* – the exposed surface opposite to that upon which active semiconductor devices have been or will be fabricated.
- *Median surface* – the locus of points in the wafer equidistant between the front and back surfaces, as shown in Figure 1.5.

Figure 1.5 shows a wafer, spinning along its centroidal axis as it passes through the gap between a pair of non-contact probes A and B, typically capacitive sensors. The distances a and b are measured by the probe sensors, with the gap distance D calibrated. As a result, the thickness of the wafer, t, at the time of the sampling and measurement as the wafer surface passes through the gap between the probes is

$$t = D - (a + b). \tag{1.4}$$

Based on the schematic of measurement in Figure 1.5 and Equation (1.4), the following bulk properties are presented.

Among the sampled points on the wafer after completion of the spin and scan, the *total thickness variation (TTV)* is calculated as the difference between the maximum and minimum values of the thickness of the wafer, as follows

$$\text{TTV} = t_{\max} - t_{\min}. \tag{1.5}$$

Further reference of TTV can be found in [ASTM Standard F1530 (2002); ASTM Standard F533 (2002); SEMI MF1530 (2018); SEMI MF533 (2010); SEMI MF657 (2014)].

Bow is the deviation of the center point of the median surface of a free, unclamped wafer from a median-surface reference plane established by three points equally spaced on a circle with diameter a specified amount less than the nominal diameter of the wafer [ASTM Standard F534 (2002); SEMI MF534 (2014); SEMI MF1390 (2018)]. Referring to Figure 1.5 with the definition of the distance z from the reference plane at the midspan between the probes A and B, we have

$$z = \frac{D}{2} - a - \frac{t}{2} \qquad \text{and} \qquad z = b + \frac{t}{2} - \frac{D}{2}.$$

Consequently, the bow of the wafer, z, can be derived from both equations above as

$$\text{bow} = z = \frac{b - a}{2}. \tag{1.6}$$

Bow is a measure of the concave or convex deformation of a wafer, based on the center of the wafer, independent of any thickness variation. Positive values of bow denote a convex (dome-shaped) median surface when the wafer is positioned with its front surface up. Conversely, negative values of bow denote a concave (bowl-shaped) median surface.

Warp is the difference between the maximum and the minimum distances of the median surface of a free, unclamped wafer from the reference plane [ASTM Standard F1390 (2002); ASTM Standard F657 (1999); SEMI MF1390 (2018); SEMI MF657 (2014)]. Like bow, warp is the measurement of the difference of the median surface with respect to a reference plane. Unlike bow, which measures only the difference at the center point of a wafer, warp uses the entire median surface of the wafer to determine the difference between the maximum and minimum distances, taking into consideration the signs of such distances. This is illustrated in Figure 1.6, in which a reference plane is established, as shown, and the maximum and minimum distances of the median surface of the wafer from the reference plane are labeled as $d_{\max}$ and $d_{\min}$, respectively. Thus, warp is obtained as

$$\text{warp} = d_{\max} - d_{\min}. \tag{1.7}$$

Note that the signs of the distances are included to determine the optimal values, and to obtain the warp. When the reference plane is at the midspan between probes A and B, as in Figure 1.5, the distance d is the same as z; thus, the warp can be deduced as follows

$$\text{warp} = z_{\max} - z_{\min} = \frac{(b - a)_{\max}}{2} - \frac{(b - a)_{\min}}{2}. \tag{1.8}$$

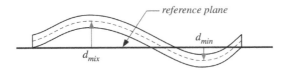

Figure 1.6 Illustration of the calculation of the warp of a wafer with respect to a reference plane. The distance d is measured from the reference plane. In the case when the reference plane is at the midspan between the two probes A and B, as in Figure 1.5, the distance d is the same as z.

Note that the warp defined in Equation (1.8), with the reference plane defined as the midspan between the two probes A and B, is only a function of the measurements a and b. However, the warp is dependent upon the location of the reference plane if defined arbitrarily.

Example 1.4.1

An example to illustrate the calculation of the values of TTV, bow and warp is shown in Figure 1.7. The reference plane is taken at the midspan between between the probes A and B, as shown in Figure 1.7.

Solution:

The TTV of this wafer according to Equation (1.5) is

$$TTV = 2 - 2 = 0$$

because the thickness happens to be the same throughout the wafer surface in this example. Normally, a non-zero TTV is expected.

The bow of the wafer, z, is measured at the center point of the wafer, and defined in Equation (1.6). It is

$$\text{bow} = \frac{(b - a)_{\text{center}}}{2} = \frac{5.3 - 4.7}{2} = 0.3.$$

Since the bow is positive, the wafer is above the reference plane in a convex shape around the center. Note that the value of bow depends on the location of the reference plane. The measurements of bow, z, at other points on the wafer are listed in the following table.

bow $z =$	-1	2	0.3	-2	-4	-1

The warp of this wafer can be determined by employing Equation (1.7). Since $d_{\text{max}} = 4$ and $d_{\text{min}} = -8$, the warp is

$$\text{warp} = \frac{(b - a)_{\text{max}} - (b - a)_{\text{min}}}{2} = \frac{4 - (-8)}{2} = 6.$$

Note that the value of warp is independent of the probe position relative to the wafer. However, change of the reference plane with respect to the wafer will affect the value of warp. Therefore, the TTV, bow and warp of the point at the center, shown in Figure 1.7, is:

$$TTV = 0, \qquad \text{bow} = 0.3, \qquad \text{warp} = 6.$$

If probe B in Figure 1.7 is shifted downward by four units, the measurements of the probe B, b, will be added to by four units. In this case, the reference plane at the halfway point between the two probes will be at the level aligned with the back side of the wafer at the leftmost point. The values at each point can be recalculated and listed in Table 1.4.

Thus, the parameters of surface quality are

$$TTV = 0, \qquad \text{bow} = 2.3, \qquad \text{warp} = 6.$$

As expected, the values of TTV and warp do not change, while the value of bow changes. The change in the bow values is exactly two units, half of the shift in b, as expected.

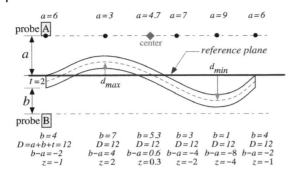

Figure 1.7 Example illustrating the calculation of TTV, bow and warp of a wafer, with the reference at the midspan between the two probes.

Table 1.4 Results of example in Figure 1.7 with probe B shifted downward by four units.

$a =$	6	3	4.7	7	9	6
$b =$	8	11	9.3	7	5	8
$D =$	16	16	16	16	16	16
$(b - a) =$	2	8	4.6	0	−4	2
$z =$	1	4	2.3	0	−2	1

1.4.2 Discussion on Warp

As observed in the preceding example, the values of bow are affected by the position of the reference plane. In the standards [ASTM Standard F657 (1999); SEMI MF657 (2014)], a wafer is supported by three hemispherical points on a reference ring that establishes a reference plane. The gravity-induced deflection will also affect the values of bow and warp. The standards in [ASTM Standard F1390 (2002); SEMI MF1390 (2018)] ratified such problems by defining warp with automated scanning. In the standards, the median surface is mathematically corrected for gravitational effects and for the mechanical signature of the instrument. A parameter, z, is defined as the distance between the wafer median surface and the point halfway, at the midspan, between the upper and lower probes. See also Figure 1.5.

The position, z, of the median surface of the wafer at each point with respect to a plane halfway between the upper and lower probes is, as discussed earlier,

$$z = \frac{b - a}{2}$$

which is the same as Equation (1.6). Gravitational and other compensations are applied to the z-position of the median surface. A reference plane is then constructed by a least-squares fit to the median surface z-position data at all points of the scan pattern. The z-value of the reference plane is called z_{ref}. The reference plane deviation (RPD) is defined as the difference between the measured z-position, z_{com}, and the reference, z_{ref}, at each point

$$\text{RPD} = z_{com} - z_{ref}. \tag{1.9}$$

A positive RPD at the center renders a dome-shaped (or convex) wafer; while a negative RPD at the center renders a bowl-shaped (or concave) wafer.

The warp is then defined as the difference between the largest and smallest of the RPD defined in Equation (1.9), by taking into consideration the signs of values,

$$\text{warp} = \text{RPD}_{\text{max}} - \text{RPD}_{\text{min}}. \tag{1.10}$$

Warp is akin to the potato-chip effect and may be caused by residual stress internally or unequal stresses on the two exposed surfaces of the wafer, or aggravated by the gravity-induced deformation and dynamics with inertia force of spinning the wafer for measurement. In recent years, different equipment and techniques have been developed, such as non-dynamic whole-surface metrology using the optical method [Wei and Kao (1999); Wei et al. (1998b)].

1.4.3 Automated Measurements of TTV, Warp, Bow, and Flatness

Note that the measurements of TTV, warp, bow, and flatness are performed by an automated process using equipment and apparatus with different means of sensing and data analysis nowadays in industry. For example, equipment performing such automated measurements will sample many points on the surface of a wafer using capacitive means of sensing and data collection in order to to calculate the TTV, warp, bow, and flatness. Various documents are published by the SEMI (https://www.semiviews.org) that describe in detail the "test methods," "test methods for measuring," and "guide for measuring" the TTV, warp, bow, and flatness. Readers are encouraged to read the reference cited in Section 1.4.1.

1.4.4 Wafer Flatness

Planarization is the process of increasing the flatness or planarity of the wafer surface. Planarization techniques can be broken down into two categories: (i) global planarization, and (ii) local planarization. The former consists of techniques that decrease long-range variations in wafer surface topology, especially those that occur over the entire image field of the stepper in microfabrication. The latter refers to techniques that increase wafer flatness over local and short distances. There are several planarization techniques used in wafer fabrication today, including but not limited to: oxidation; chemical etching; taper control by ion implant damage; deposition of films of low-melting point glass; sacrificial wafer bonding; resputtering of deposited films for smoothing; use of polyimide films; use of new resins and low-viscosity liquid epoxies; use of spin-on glass (SOG) materials; sacrificial etch-back; and chemical-mechanical polishing (CMP).

The starting prime wafers for semiconductor device fabrication are flat or planar within the specification. However, layers of materials with different shapes and depths are deposited onto the wafer surface as wafers undergo various steps in device fabrication, through continuous deposition and removal processes. These repeated deposition and removal steps cause wafers to lose their flatness or planarity. As the number of layers and interconnect techniques for IC fabrication evolves, the planarization of the wafer surface is further aggravated.

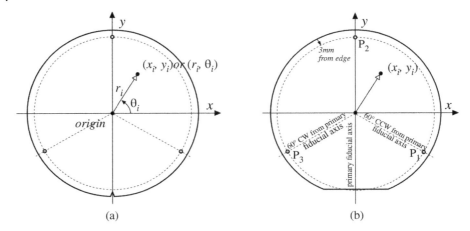

Figure 1.8 Definition of the planar coordinate system of wafers: (a) on a wafer with a notch orientation having a point on the wafer identified with the Cartesian coordinates, (x_i, y_i), or polar coordinates, (r_i, θ_i); (b) the same coordinate definition on a wafer with a primary orientation flat. In addition, three equally spaced points, P_1, P_2 and P_3, are illustrated with respect to the primary fiducial axis perpendicular to the orientation flat, or notch, along the y axis, on a circle 3 mm away from the edge of wafer for constructing the three-point reference plane.

A decrease in the flatness of the wafer's surface introduces problems during device fabrication. Maintaining continuity of fine lines without breaks becomes more difficult when the wafer surface flatness is decreased. In addition, progressive loss of planarity eventually makes the imaging of fine-featured patterns on the wafer increasingly challenging.

The standards of wafer flatness and measurements were described in [ASTM Standard F1530 (2002); ASTM Standard F533 (2002)], and in SEMI document [SEMI 3D12 (2015); SEMI MF533 (2010); SEMI MF1530 (2018)]. The following presentation will follow the standards in SEMI [SEMI MF1530 (2018)] for determining wafer flatness. It is also noted that the measurement of wafer flatness is now done using an automated process and equipment, as described in Section 1.4.3. The following procedures are meant to provide an understanding of how flatness is determined.

Equation (1.4) can be re-written as follows to express the data set of thickness measurements, $t(x, y)$, of many sampled points on the surface of a wafer with planar coordinates x and y

$$t(x, y) = D - [a(x, y) + b(x, y)] \tag{1.11}$$

where $x = y = 0$ is the origin of the planar coordinate system at the center of the wafer. A definition of the coordinate system is illustrated in Figure 1.8. The definition of the parameters in Equation (1.11) are the same as those presented in Section 1.4.1.

Flatness Determination

The flatness determination is performed using the following steps [SEMI MF1530 (2018)].

1. Construct a reference plane by using the following equation

$$Z_{\text{ref}} = a_{\text{R}} x + b_{\text{R}} y + c_{\text{R}} \tag{1.12}$$

where a_R and b_R are the slopes of the reference plane in the x and y directions, respectively, and c_R is the distance of the origin of the reference plane from the back surface. The coefficients a_R, b_R and c_R are determined in the following three ways.

(a) If the reference plane is the "ideal back surface," then the coefficients are
$a_R = b_R = c_R = 0$.

(b) If the reference plane is the "least-squares reference plane," the coefficients are chosen such that the following objective function is minimized over the fixed quality area (FQA)

$$\sum_{x,y} [t(x,y) - (a_R\, x + b_R\, y + c_R)]^2.$$

The FQA for consideration can be global or site or subsite determination.

(c) If the reference plane is a "three-point reference plane," this plane is formed by three equally spaced points, $P_1(x_1,y_1)$, $P_2(x_2,y_2)$ and $P_3(x_3,y_3)$, located on a circle whose perimeter is 3 mm from the edge of a wafer of nominal diameter, as illustrated in Figure 1.8, using the following three equations

$$\begin{cases} t(x_1,y_1) = a_R\, x_1 + b_R\, y_1 + c_R \\ t(x_2,y_2) = a_R\, x_2 + b_R\, y_2 + c_R. \\ t(x_3,y_3) = a_R\, x_3 + b_R\, y_3 + c_R \end{cases}$$

2. Construct a focal plane, parallel to the reference plane to calculate deviation parameters from the data set of thickness measurements $t(x,y)$, as follows

$$Z_{focal} = a_F\, x + b_F\, y + c_F \tag{1.13}$$

where $a_R = a_F$ and $b_R = b_F$ are required in Equations (1.12) and (1.13) because the reference plane and the focal plane are parallel to each other, with the same slopes in both x and y directions. A global focal plane is the same as the corresponding reference plane with $c_F = c_R$. A site or subsite focal plane is displaced from the corresponding reference plane, with the coefficient determined by the following equation

$$c_F = t(x_0,y_0) - (a_F\, x_0 + b_F\, y_0)$$

where (x_0, y_0) are the coordinates of the centroid of the site or subsite.

3. Determine the point-by-point differences between the thickness and the reference or focal plane by

$$f(x,y) = t(x,y) - (ax + by + c) \tag{1.14}$$

where the coefficients a, b and c are replaced by the respective coefficients from Equations (1.12) and (1.13) based on the plane used.

4. Determine the range (also called the "total indicated reading", TIR) which is defined as

$$\text{TIR} = f(x,y)|_{max} - f(x,y)|_{min} \tag{1.15}$$

where $f(x,y)$ is the function defined in Equation (1.14), $f(x,y)|_{max}$ is the largest (most positive) value of $f(x,y)$ over the range of x and y within the FQA, and $f(x,y)|_{min}$ is the smallest (most negative) value of $f(x,y)$ over the range of x and y

within the FQA. It is noted that the same results of TIR are obtained by employing either the reference plane or the focal plane.

5. Determine the focal plane deviation (FPD) which is the larger of the absolute values of $f(x, y)|_{max}$ or $f(x, y)|_{min}$ with a positive or negative sign.

$$\begin{cases} FPD = + \parallel f(x,y)|_{max} \parallel, & \text{if } \parallel f(x,y)|_{max} \parallel > \parallel f(x,y)|_{min} \parallel \\ FPD = - \parallel f(x,y)|_{min} \parallel, & \text{if } \parallel f(x,y)|_{max} \parallel < \parallel f(x,y)|_{min} \parallel \end{cases}$$

where $\parallel \cdot \parallel$ denotes the absolute value.

Table 1.5 tabulates the acronyms to qualify the wafer flatness measurement based on the preceding procedures described [SEMI MF1530 (2018)]. In the following, a few examples are used to explain the designation of the individual characters in the acronyms based on the table and the preceding procedures.

- GBIR is a **g**lobal measurement, with the **b**ack surface of the wafer as the reference surface, with an **i**deal back wafer surface on the entire FQA as the reference plane, and with the flatness measurement in **r**ange, or TI**R**.
- SBID is a **s**ite measurement, with the **b**ack surface of the wafer as the reference surface, with an **i**deal back surface on the entire FQA as the reference plane, and with the flatness measurement in FP**D**.
- SF3R is a **s**ite measurement, with the **f**ront surface of the wafer as the reference surface, with a **3**/three-point reference plane, and with the flatness measurement in in **r**ange, or TI**R**.
- SFQR is a **s**ite measurement, with the **f**ront surface of the wafer as the reference surface, with the least-s**q**uares reference plane, and with the flatness measurement in **r**ange, or TI**R**.
- SFSR is a **s**ite measurement, with the **f**ront surface of the wafer as the reference surface, with the least-squares reference plane on sub**s**ite, and with the flatness measurement in **r**ange, or TI**R**.

Example 1.4.2 The "ideal back surface" of wafer is used as the reference plane for the flatness determination. Explain and derive the coefficients of the equations for flatness determination for GBIR.

GBIR has a reference plane of the "ideal back surface." Thus, the coefficients in Equation (1.12) are $a_R = b_R = c_R = 0$. This confirms that $Z_{ref} = 0$, and the reference plane is the back surface. With a global focal plane, we have $c_F = c_R = 0$. Now Equation (1.14) reduces to $f(x, y) = t(x, y)$ – the thickness of the wafer. The TIR Equation (1.15) is the same as the difference between the maximum and minimum thicknesses of the wafer, which is the definition of the total thickness variation, or the TTV. Therefore, GBIR is the same as TTV. Nevertheless, TTV can be obtained from the $t(x, y)$ data over the entire FQA, without having to construct the reference plane and going through the procedures in this section. Historically, TTV has been obtained before the protocol of the flatness determination. ∎

In Example 1.4.2, it is ascertained that GBIR is the same as TTV. The different measurement methods of wafer flatness in Table 1.5 enables quantitative determination of wafer flatness both globally and locally. The reference plane can be constructed

Table 1.5 Four-character acronyms for wafer flatness measurement.

Acronym	Measurement method	Reference surface	Reference plane	Flatness measurement
GBIR	Global	Back	Ideal back surface, on entire FQA	TIR
GF3R	Global	Front	Three-point	TIR
GF3D	Global	Front	Three-point	FPD
GFLR	Global	Front	Least-squares, on entire FQA	TIR
GFLD	Global	Front	Least-squares, on entire FQA	FPD
SBIR	Site	Back	Ideal back surface, on entire FQA	TIR
SBID	Site	Back	Ideal back surface, on entire FQA	FPD
SF3R	Site	Front	Three-point	TIR
SF3D	Site	Front	Three-point	FPD
SFLR	Site	Front	Least-squares, on entire FQA	TIR
SFLD	Site	Front	Least-squares, on entire FQA	FPD
SFQR	Site	Front	Least-squares, site	TIR
SFQD	Site	Front	Least-squares, site	FPD
SFSR	Site	Front	Least-squares, subsite	TIR
SFSD	Site	Front	Least-squares, subsite	FPD

Notes: The four characters of the acronyms represent each of the four columns in this table, with (1) measurement method, (2) reference surface, (3) reference plane, and (4) flatness measurement parameters. The flatness measurement parameters are either "TIR – the range" (in terms of total indicated reading, TIR) or "FPD – the deviation" (focal plane deviation, FPD). The measurement method can encompass the entire wafer with "global" measurement or local "site."

using ideal back surface, least-squares best fit, of using three equally spaced points on surface.

Readers may want to reference [SEMI MF1530 (2018)] for more details of other scenarios of flatness determination.

1.4.5 Nanotopography or Nanotopology

Nanotopography or nanotopology refers to the variation of height of the surface of wafers over specified distances, which must be controlled to meet the requirements of microelectronics fabrication process steps. Parameters pertaining to the nanotopography have been included in the International Technology Roadmap for Semiconductors (ITRS) since 2003 [Allan et al. (2002); Edenfeld et al. (2004)]. Both front-surface and back-surface topography will impact wafer suitability for selected process steps. SEMI M43 (2018) defines nanotopograpy as the "non-linear deviation of the whole front wafer surface within a spatial wavelength range of approximately 0.2 to 20 mm and within the fixed quality area (FQA)." A nanotopography reporting flow is introduced in this SEMI guide. The guide for determining nanotopography of unpatterned silicon wafers for the 130 nm to 22 nm generations in high volume manufacturing is specified in [SEMI M78 (2018)], with procedures and a decision tree provided.

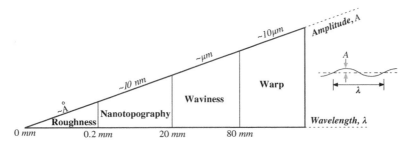

Figure 1.9 Classification of topographical features based on their spatial wavelengths. The order of amplitude of each feature is also illustrated. Source: Bhagavat et al. (2010). © 2010 Elsevier.

The machining processes employed to shape wafers, such as slicing, lapping, and polishing cause the surface of the wafer to assume certain shapes and characteristics. Much research effort has been invested in abrasive machining processes to understand the effect of different process parameters such as slurry concentration, abrasive size distribution, applied loads on material removal rate, surface roughness of the work piece, material removal mechanism, and effect of material properties .

The nature of such processes and machine tools can lead to non-uniformity over the wafer surface, resulting in roughness, waviness and warp, with topographical features. The topographical features are typically classified based on their spatial wavelength, as illustrated in the nanotopography chart in Figure 1.9. In special cases, the roughness and flatness of wafers may not represent the actual surface nanotopography of concern. The definition of nanotopography can depict the surface of wafers more meticulously without ambiguity in certain special cases. For example, nanotopography is defined as the deviation of a surface with a spatial wavelength of 0.2 to 20 mm, which has an amplitude of 10 nm after the final polishing process [SEMI; Bhagavat et al. (2010)].

1.4.6 Surface Roughness

Surfaces are characterized by surface texture and surface integrity. Surface texture refers to the topology or geometry of the surface and is most commonly represented by surface roughness. Surface integrity is concerned with the material characteristics immediately beneath the surface and the changes in subsurface features, especially after manufacturing and/or fabrication processes. Surface roughness is a measure of the surface texture – a topic of concern in many areas, such as tribology, design, and manufacturing. Roughness quantifies the deviation of a surface from its ideal form, often describes small, finely spaced deviation from the nominal or ideal surface. Roughness is expressed with respect to its height, width, and distance along the surface. Roughness is different from waviness, which is a recurring deviation (wave) from a reference surface. It is measured as distances of the repeating patterns of wave on the surface, such as spatial frequency (periodic distance on the surface) of the wave, waviness height and width.

Arithmetic Mean Roughness, R_a

A surface profile is illustrated in Figure 1.10, where a section of the surface from point A to point B in linear dimension of x is plotted, with the corresponding variations z from the nominal surface, as shown. A point P on the surface at x_i has a variation of z_i. If the variation, $Z(x)$, is expressed as a function of the linear dimension, the conventional roughness R_a, expressed as the arithmetic mean value, is defined as

$$R_a = \frac{1}{L} \int_0^L |Z(x)| \, dx. \tag{1.16}$$

Equation (1.16) is a theoretical continuous integration over the length of the surface profile. In practice, the surface roughness, R_a, is determined by sampling and measuring the variations of N discrete points, often equally spaced, from points A to B of the surface profile. One such discrete point is point P, shown in Figure 1.10 with a variation of z_i. The discrete equivalence of equation (1.16) with the measurements of N discrete points is

$$R_a = \frac{1}{N} \sum_{i=1}^N |z_i|. \tag{1.17}$$

Equation (1.17) is the most commonly used expression of average roughness, R_a, in the literature. The linear scan to sample points between A and B can be performed by instrument such as a mechanical profilometer with a stylus.

Root-mean-square Roughness, R_q

Another roughness measure is the root-mean-square roughness, defined as

$$R_q = \sqrt{\frac{1}{L} \int_0^L [Z(x)]^2 \, dx} \tag{1.18}$$

with its discrete equivalence as

$$R_q = \sqrt{\frac{1}{N} \sum_{i=1}^N z_i^2}. \tag{1.19}$$

Equation (1.19) magnifies the variation z_i due to the squares of variations, as compared with the arithmetic mean roughness, R_a.

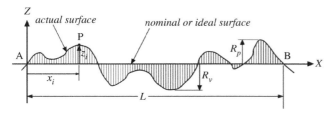

Figure 1.10 Illustration of a surface profile and roughness with the terminology for the measurements of roughness.

Peak-to-valley Roughness or Total Height Roughness, R_t

The roughness can also be quantified by the total height from peak to valley of the surface profile. This roughness is often called the *total height* or *peak-to-valley* roughness, denoted as R_t. Let's define the maximum peak height as $R_p = \max_i(z_i)$ and the maximum valley depth as $R_v = |\min_i(z_i)|$. Note the absolute sign on R_v because the value is negative without the $|\cdot|$ operation. Thus, the peak-to-valley roughness or total height roughness is given by

$$R_t = R_p + R_v = \max_i(z_i) + |\min_i(z_i)|. \tag{1.20}$$

Figure 1.10 illustrates R_p and R_v for the calculation of the roughness R_t in Equation (1.20). The R_t roughness, also discussed in ASME B46.1 and ISO 4287, is essentially the TTV of a linear scan of surface in a range of L.

Mean Roughness Height or Average Maximum Height, R_z

Figure 1.11 illustrates the calculation of the roughness measure R_z, called the *mean roughness height* or the *average maximum height*. The roughness R_z is defined as the average of the successive values of R_{ti}, each calculated over an evaluation region of l within the total length L, as shown in Figure 1.11. The parameter R_z is the same as R_z(DIN) when there are five (5) sampling regions, l, within the evaluation length of L. The mean roughness height R_z is defined as follows

$$R_z = \frac{1}{n}\sum_{i=1}^{n}(R_{ti}) \tag{1.21}$$

where $R_z = R_z$(DIN) when $n = 5$.

Ten-point Roughness Height or Ten-point Height of Irregularities, R_z(ISO)

Another definition of R_z by ISO is the *ten-point roughness height* or the *ten-point height of irregularities*, denoted as R_z(ISO). The definition of R_z(ISO) is best illustrated in Figure 1.12. The five maximum peaks and valleys within the evaluation length L are identified with magnitude in descending order by $P_1 \cdots P_5$ and $V_1 \cdots V_5$, respectively. The heights of $h_1 \cdots h_5$ are the heights between the peaks and valleys of the corresponding pair of points, as illustrated in Figure 1.12. The R_z(ISO) is thus defined as the average of the heights of the five maximum peaks and valleys, $h_1 \cdots h_5$, as in the following equation

$$R_z(\text{ISO}) = \frac{1}{5}\sum_{i=1}^{5}(h_i). \tag{1.22}$$

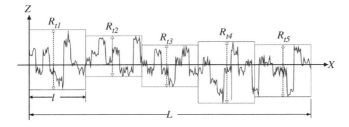

Figure 1.11 Illustration of the mean roughness height R_z, or the average maximum height, measured over five sampling regions, l, of a total evaluation length L.

Figure 1.12 Illustration of the ten-point roughness height R_z(ISO), or the ten-point height of irregularities, measured over a sampling length of L the first five largest peaks and valleys.

Table 1.6 summarizes the various roughness measures in the preceding presentation. Reference materials of various roughness measures can be found, such as those in [ASME B46 (2009); ISO-4287 (1996); Mitutoyo (2016); Precision Device (2016)].

The application of measures of surface roughness, especially the root-mean-square roughness, was defined by SEMI [SEMI MF1811 (2016); SEMI M40 (2014)]. SEMI MF1811 (2016) provides information on the extraction of the roughness parameters from linear scans of the surface of polished wafers. A mean-square roughness, R_q^2, was defined by SEMI, with the equivalent formulation of the root-mean-square roughness as

$$R_q = \sqrt{\lim_{L \to \infty} \frac{1}{L} \int_{-L/2}^{L/2} [Z_d(x)]^2 \, dx} = \sqrt{\int_0^\infty S_1(f_x) \, df_x} \qquad (1.23)$$

where $Z_d(x)$ is the "*detrended surface profile*"[2], equivalent to $Z(x)$ in equation (1.18), and $S_1(f_x)$ is the power spectral density function. In addition to R_q, the mean-square slope was also defined as the average value of the square of the slope

Table 1.6 A summary of various surface roughness measures.

Arithmetic mean roughness	$R_a = \frac{1}{N} \sum_{i=1}^{N} |z_i|$
Root-mean-square roughness	$R_q = \sqrt{\frac{1}{N} \sum_{i=1}^{N} z_i^2}$
Peak-to-valley roughness	$R_t = \max_i(z_i) + |\min_i(z_i)|$
Mean roughness height	$R_z(\text{DIN}) = \frac{1}{n} \sum_{i=1}^{n} (R_{ti})$
Ten-point roughness height	$R_z(\text{ISO}) = \frac{1}{5} \sum_{i=1}^{5} (h_i)$

———

2 The detrended profile of surface is represented by the raw or measured profile after removing instrumental and surface trends. It is noted that the rms roughness, R_q, is distinct from the average roughness, R_a, except when the two are related through a specific height-distribution function. For example, when such distribution function is Gaussian, they are related by the equation $R_a = \sqrt{\frac{2}{\pi}} R_q = 0.798 R_q$ [SEMI MF1811 (2016)].

of the detrended profile,

$$\Delta_q^2 = \lim_{L \to \infty} \frac{1}{L} \int_{-L/2}^{L/2} \left[\frac{dZ_d(x)}{dx} \right]^2 dx = \int_0^\infty (2\pi f_x)^2 S_1(f_x) \, df_x. \tag{1.24}$$

SEMI defines a seven-step procedure for roughness measurement specification of planar surfaces on polished wafers [SEMI M40 (2014)]. The procedure is enumerated in the following:

1. Select the type of instrument, including three main types as follows:
   (a) Profilometer: including atomic force microscope (AFM), other scanning probe microscopes (SPM), optical profilometer (OPR), and mechanical stylus (MPR).
   (b) Interferometer: including interference microscope (IM).
   (c) Scatterometer: including total integrating scatterometer (TIS), angle-resolved light scatterometer (ARLS), and scanning surface inspection system (SSIS).
2. Select the measurement pattern, including center point (1), five-point (5), nine-point (9), full-FQA raster scan (R), concentric full-FQA R-theta scan (C), and spiral R-theta scan (S); see [SEMI M40 (2014)] for more details with illustration.
3. Select the pattern orientation: including type A (A) and type B (B); see [SEMI M40 (2014)] for more details with illustration.
4. Select the local measuring condition: including point (P), line (L), or area (A).
5. Select the parameters to be determined: including arithmetic average roughness, R_a (A), root-mean-square roughness, R_q (Q), ten point roughness, R_z-ISO (Z), or peak to valley, R_t (T).
6. Specify the measurement calculations to be reported: including average (A), range (R), maximum (M) or one standard deviation, one sigma (D).
7. Specify the bandwidth and scan length limits within which data is to be gathered.
8. Finally, record the abbreviation associated with these selections as described above, separating adjacent abbreviations with a comma and using periods for decimal notation. This creates a seven-field abbreviation that follows the steps 1 to 7 of this procedure.

Example 1.4.3 An example of roughness measurement specifications and related output is used to illustrate the SEMI procedures. A mechanical stylus profilometer measurement was specified as MPR,9,A,L,A,A,250/15 with an output measurement of $R_a = 0.53$ nm.

Based on the seven-step procedure outlined by SEMI, the elements and codes are listed listed in the following, corresponding to the seven steps

Elements:	(1)	(2)	(3)	(4)	(5)	(6)	(7)
	Instrument	Pattern	Orientation	Local	Parameter	Calculation	Bandwidth
	mech profiler	nine-point	type	line scan	R_a	average	250/15 µm
Codes:	MPR	9	A	L	A	A	250/15

The output example is: MPR,9,A,L,A,A,250/15 $= 0.53$ µm, having an arithmetic average roughness measure of 0.53 µm with the specification of measurements and related output. ∎

1.5 Other Properties and Quality Requirements of Silicon Wafers

In addition to the wafer surface measurements presented in Section 1.4, several other properties of silicon wafers are discussed in the following sections.

1.5.1 Mechanical and Material Properties

Silicon is an anisotropic crystalline material whose properties depend on the orientation relative to the crystal lattice. Consequently, the elastic or Young's modulus (E) and the Poisson ratio (v) need to be considered as a function of crystalline orientation for the analysis of mechanical and material behavior.

Values of the material properties, E and v, based on the study by Hopcroft et al. (2010); Wortman and Evans (1965); Bhagavat and Kao (2006) are listed in Table 1.7.

1.5.2 Property of Silicon with Anisotropy

Three orientations of silicon crystal are most commonly used: they are (100), (110) and (111). For each type of the three silicon crystals and wafers, the material properties are not isotropic. That is, when one traverses the surface of a silicon wafer, one will observe different material properties by approaching from different directions. Before we present such anisotropic properties of modulus of elasticity, it will help to define the "direction of approach" or DOA. The DOA is defined as the direction from which one traverses along the surface of a single-crystalline wafer with a specific crystal orientation. As in Table 1.3, the Miller index is used here to denote plane with (…), and direction with […].

A 3D example is illustrated in Figure 1.13 to aid the understanding of the terminology. Figure 1.13 illustrates a (100) ingot and wafer surface. First, we recognize that any direction lying in (100) plane qualifies as a DOA. Figure 1.13 illustrates the case where [010] is chosen as the DOA. Different DOAs can be chosen by rotating

Figure 1.13 3D illustration for a DOA in (100) crystal. The illustrated DOA is in the [010] direction, corresponding to the angle $\theta = 0°$ in Figure 1.14. Source: Bhagavat and Kao (2006). © 2006 Elsevier.

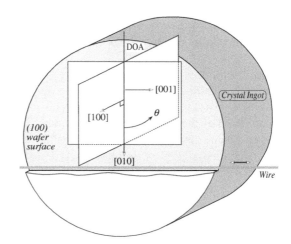

Table 1.7 Material properties of the Young's modulus and Poisson's ratio of silicon.

Young's modulus, E	Poisson's ratio, v	Density
130 to 188 GPa	0.048 to 0.4	2.329 g cm^{-3}

Sources: Bhagavat S and Kao I 2006 Theoretical analysis on the effects of crystal anisotropy on wiresawing process and application to wafer slicing. International Journal of Machine Tools and Manufacture 46, 531–541; Hopcroft M A, Nix W D and Kenny T W 2010 What is the young's modulus of silicon?. J. of Microelectromechanical Systems 19(2), 229–238 and Wortman J J and Evans R A 1965 Young's modulus, shear modulus, and poisson's ratio in silicon and germanium. J. Appl. Phys. 36(1), 153–156.

the ingot or wafer in Figure 1.13 in a clockwise sense with respect to the [100] direction. This DOA is a very helpful concept in orienting the crystal when the ingot is sliced into wafers. See also Chapter 2.

Graphical illustrations of Young's modulus of silicon with the variation of direction curves with different DOAs for these three orientations of crystal are presented in the following.

(100) Crystal Wafer

Following the terminology and definition of DOA in Figure 1.13, the DOA is rotated in a clockwise sense with the angles of $\theta = 0°, 15°, 30°, 45°, 60°, 75°, 90°$ from the [010] direction towards the [001] direction, as depicted in Figure 1.14(a), in increments of $\theta = 15°$. With each increment of angle θ, a corresponding direction curve in the plane determined by the corresponding DOA and the wafer surface normal [100], is drawn in Figure 1.14(b) using the same legend used for depicting the corresponding DOA in Figure 1.14(a). For example, when the DOA is at $\theta = 15°$ from the [010] direction, indicated with an arrow of "*" in Figure 1.14(a), the corresponding direction curve is the one connected with the legend "*" in Figure 1.14(b).

It is noted from the results plotted in Figure 1.14 that the pattern reverses as θ moves from 45° to 90°; that is, the direction curve for a DOA at an angle of $(45° + \alpha)$ from [010] is the same as the direction curve for the DOA at an angle of $(45° - \alpha)$ from [010], where $0° \leq \alpha \leq 45°$. The pattern obtained by varying the DOA in the first quadrant, as shown in Figure 1.14, is repeated over the next three quadrants due to the nature of the crystal symmetry of silicon.

(110) Crystal Wafer

Similarly, Figure 1.15(a) illustrates the variation of DOA with respect to the [001] direction on a (110) wafer. Figure 1.15(b) illustrates the direction curve in the plane determined by the corresponding DOA and wafer surface normal [110]. The figure shows the direction curves for the DOAs within an angle of 180° of [001]. The pattern reverses as we go from 90° to 180°; that is, the direction curve for a DOA at an angle of $(90° + \alpha)$ from [001] is the same as the direction curve for the DOA at an angle of $(90° - \alpha)$ from [001], where $0° \leq \alpha \leq 90°$. The pattern obtained by varying the DOA in the first two quadrants, as shown in Figure 1.15, is repeated over the next two quadrants due to the nature of the crystal symmetry of silicon.

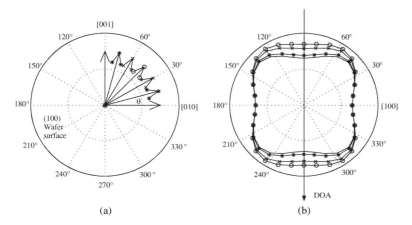

(a)   (b)

Figure 1.14 (a) Variation of DOA with respect to the [010] on (100) crystal. As the ingot shown in Figure 1.13 rotates clockwise in increments of 15°, the DOA changes from [010] towards [001] as shown. (b) Direction curve for the Young's modulus in the plane formed by the DOA and wafer surface normal [100]. The pattern observed in the first quadrant is illustrated. The legend used to indicate the DOA and the corresponding curve is the same. DOAs with the same curves are illustrated using the same legend. The pattern repeats itself over the next three quadrants due to the nature of the crystal symmetry of silicon. Source: Bhagavat and Kao (2006). © 2006 Elsevier.

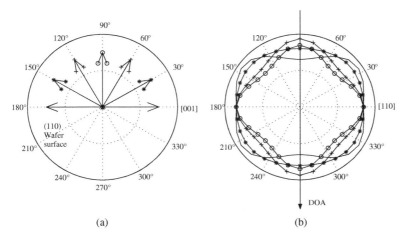

(a)   (b)

Figure 1.15 (a) Variation of DOA with respect to the [001] direction on (110) crystal. (b) Direction curve for Young's modulus in the plane formed by the DOA and wafer surface normal [110]. The pattern observed in the upper half plane is illustrated. The legend used to indicate the DOA and the corresponding curve is the same. DOAs with the same curves are illustrated using the same legend. The pattern repeats itself over the lower half plane due to the nature of crystal symmetry of silicon. Source: Bhagavat and Kao (2006). © 2006 Elsevier.

1.5.3 Gravity-induced Deflection of Wafers

The material properties presented in the preceding sections can be used to determine the gravity-induced deflection (or self-weight deflection) of silicon wafers during handling and processing. The thickness of the wafer in Table 1.2 is chosen for each size so that the wafers will not have too much deflection in processing and fabrication, and so that the wafers will not break during handling because of its own weight. A finite element analysis (FEM) of 300 mm wafers can be performed to estimate the deflection and stress of the wafer due to its own weight. In the following, two cases of analysis are considered:

1. Bending deflection due to its own weight with one side supported (one axis of symmetry)
2. Deflection due to its own weight with three-point support

For the analysis here, a (100) wafer having a diameter of 300 mm and a thickness of 775 μm is considered with a Young's modulus $E(100) = 130$ *GPa* and a Poisson's ratio of $v = 0.048$. When a (111) wafer is used under similar conditions, a Young's modulus of $E(111) = 188$ *GPa* and a Poisson's ratio of $v = 0.4$ are employed in the analysis. The density of silicon is $\rho = 2.329$ g cm^{-3}.

- **Case I: Deflection with one side supported.** In this case, the analysis is performed using a (100) wafer with a diameter of 300 mm and a thickness of 775 μm, constrained on the top side, as shown in the figures. The material properties are $E(100) = 130$ GPa with a Poisson's ratio of $v = 0.048$. The wafer is clamped along the X-axis, which is aligned with the [110] directions – the direction indicated by the notch on the wafer. The gravity is along the Z-direction, causing deflection in the same direction due to self-weight.
  The results of FEM are shown in Figure 1.16. The maximum deflection of a (100) wafer is about 3.6 mm at the free end of the wafer. Figure 1.16 illustrates the deflection pattern and magnitudes. If a (111) wafer is used under the same boundary condition, the maximum deflection is 1.87 mm, due to a different Young's modulus E.
- **Case II: Deflection with three-point support.** In this case, the analysis is performed using a (100) wafer with a diameter of 300 mm and a thickness of 775 μm, resting on three hemispherical supports on the back surface. Different coefficients of friction at the contacts are assumed in the simulation with values ranging from 0.1 to 0.01 with similar results.
  The results of FEM are shown in Figure 1.17. The maximum deflection of the wafer is about 61 μm at the center of the wafer. This deflection, although much smaller than the previous case, can cause the slight variation comparable to the minimum feature size of microelectronics fabrication. Proper design of constraints to support the 300 mm silicon wafers is important to increase the yield of the fabrication.

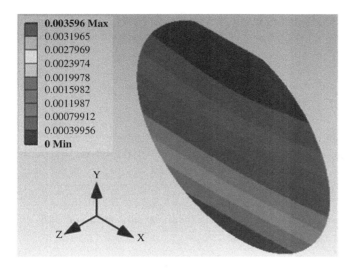

Figure 1.16 Plot of gravity-induced deflection of a 300 mm wafer due to its own weight with one side constrained using FEM analysis.

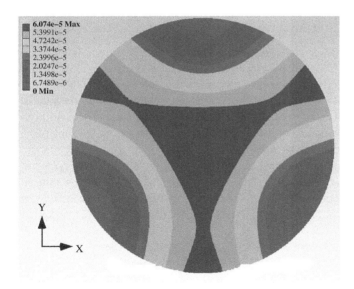

Figure 1.17 Plot of gravity-induced deflection of a 300 mm wafer due to its own weight with three-point support using FEM analysis. The maximum deflection is about 61 μm at the center of the wafer surface.

1.5.4 Wafer Edge Properties

The edge property of wafers is one of the most important properties that determines wafer durability in the wafer manufacturing and fabrication process. This is because most cracks start along edges first and spread across the wafer, destroying it through crack propagation typically found in brittle materials. Edge rounding is a step that minimizes the crack forming on the edge of the wafers in order to prevent the crack propagation that destroys wafers. See also Chapter 9.

1.6 Economics of Wafer Manufacturing

The economics of wafer manufacturing depends on the type of wafer. Silicon wafers, including semiconductor prime wafers and photovoltaic wafers, have been the most dominant type of wafer produced to date; hence, a lot of focus has been placed on the economics of producing silicon wafers than other wafers combined. In the following, the three different categories of wafers are discussed with illustration, followed by the discussion of the cost of wafers in the microelectronics industry.

1.6.1 Three Categories of Wafers

Wafers produced by the wafer manufacturing processes, as will be discussed in Chapter 2, can be broken down into three categories, based on the purpose for which it serves, as follows.

1. Semiconductor prime wafers: These wafers are used in semiconductor fabrication, made of materials such as silicon, III-V and II-VI compounds, silicon carbide, and others. Such wafers are subject to very stringent requirements of surface finish due to the rigorous microelectronics fabrication processes. In Figure 1.18, a lithium niobate ($LiNO_3$) wafer is shown on the left, and a silicon prime wafer is shown on the right. The orientation flats of both wafers can be seen clearly. Both wafers have gone through the wafer manufacturing processes and are ready for microelectronics fabrication to make devices.
2. Photovoltaic (PV) wafers: Silicon-based PV wafers are often made of polycrystalline silicon, with some made of single crystalline silicon. The silicon materials for PV wafers are often trickled down from the semiconductor industry. The wafers can be either round or rectangular, as opposed to the round semiconductor

Figure 1.18 Photos of a lithium niobate wafer (left), and a silicon prime wafer (right). Source: Imin Kao.

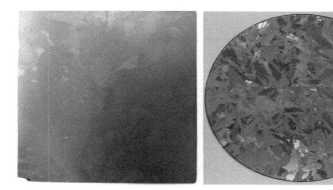

Figure 1.19 Photos of polycrystalline silicon PV wafers with rectangular and round cross sections. Rectangular PV wafers are more commonly used in the production of PV modules. Source: Imin Kao.

wafers. Usually integrated with the production of solar modules, the PV wafers do not have nearly as stringent requirements on surface finish as those of the semiconductor wafers. PV wafers as-sliced by a wiresaw can often be used to produce photovoltaic panels or modules. directly. Typical PV wafers are shown in the photos in Figure 1.19 with a square and a round polycrystalline silicon wafer. The grain boundaries can be seen visually.

3. Other wafers: Wafers in this category are often specialty wafers or products under research and development, and can be made from various materials. Figure 1.20 shows on the left a photo of a piece of aluminum sliced by a wiresaw, which is often designed and used to slice brittle materials such as silicon. The surface of the ductile aluminum material has a large surface roughness, larger than typical brittle materials sliced by a wiresaw. Figure 1.20 also shows a photo of a slice of kevlar material cut by a wiresaw on the right. Kevlar, a very strong material, is used to make bullet-proof vests and was used as sheet material to wrap the lunar module which landed on the moon. The surface of the as-sliced Kevlar is very smooth with very small surface roughness.

Figure 1.20 Photos of a thin piece of aluminum sliced by a wiresaw (left), and a slice of Kevlar cut by a slurry wiresaw (right). Source: Imin Kao.

1.6.2 Cost of Silicon Wafers

In microelectronics fabrication, "cost per wafer" is often used as a benchmark to assess the total cost of processing in order to make modification and adjustments that make the industry more profitable [Byrne et al. (2013); Leachman et al. (2007); SEMI (2012b, 2014)]. The IC manufacturing costs can be divided into three categories:

1. Materials costs, including the cost of prime wafers from the wafer manufacturing
2. Capital costs
3. Labor costs.

A large percentage of the total cost is due to the capital costs with investment in equipment and technology. Material costs and consumables are the second highest costs – including the cost of the semiconductor source materials and the cost of production of prime wafers. The labor costs are the smallest percentage because the process is highly automated and controlled. Nonetheless, semiconductor fabs have found their homes in countries with lower labor costs with highly skilled workers. Overall, the cost per wafer has risen dramatically over the last few decades, partly because of the cost of replacing high-cost equipment to process larger wafers, as well as the cost of wafer manufacturing to produce prime wafers of larger diameters.

In a study on the silicon-based solar industry using silicon wafers as a source to produce PV cells, it was estimated that the total cost of the silicon and wafering (or wafer slicing) process represents over 50% of the total cost [Applied Materials (2011)]. The additional cost of producing wafers for the semiconductor industry, as compared to the PV industry, is the cost associated with the post-slicing processes, which include lapping, polishing, and packaging.

The cost of wafer manufacturing depends on parameters such as the silicon material (as single-crystalline or polycrystalline silicon), wafers produced per kilogram of silicon material (versus the kerf loss[3]), quality and yield, consumables of the slicing process, and initial investment of equipment. The total cost of slicing per wafer can be estimated as follows

$$C_{\mathrm{w}} = \frac{C_{\mathrm{si}} + C_{\mathrm{s}} + C_{\mathrm{o}}}{N_{\mathrm{w}}} \tag{1.25}$$

where C_{w} is the total cost of slicing per wafer, C_{si} is the cost of silicon source material, C_{s} is the cost of consumables during the slicing process, such as abrasives and carrier fluid, and C_{o} encompasses other costs such as the fixed cost, equipment, and power. The parameter N_{w} is the total number of yielded wafers per cut. It is the product of the total number of expected wafers and the quality yield of the cut, as follows

$$N_{\mathrm{w}} = N \, \eta \tag{1.26}$$

where η is the quality yield per cut which includes the yield during the slicing process and the post-processing handling. The quality yield, η, is a measure of efficiency of the manufacturing process to eliminate waste produced by wafer

---

3 The kerf loss represents the loss of silicon material between two yielded wafers during the slicing process.

breakage or defect during slicing and post process handling. Certain wafer defects may be due to the growth process of ingot. The total number of expected wafers in Equation (1.26) is defined in the following

$$N = \frac{L}{t} \tag{1.27}$$

where L is the effective length of the silicon ingot spanned by the wire web of the wiresaw, and t is the sum of the thickness of the yielded wafer, t_w, and the kerf loss, t_k (see also Chapter 2); that is,

$$t = t_w + t_k. \tag{1.28}$$

The cost of slicing, as in Equation (1.25), together with the cost for lapping, polishing, and packaging will represent the total cost of wafer manufacturing, as in the following equation

$$C = C_w + C_l + C_p + C_g \tag{1.29}$$

where C is the total cost of producing each wafer, C_w is defined in Equation (1.25), and C_l, C_p, C_g are the cost for lapping, polishing, and packaging per wafer, respectively.

1.7 Summary

In this chapter, the wafers used for semiconductor, photovoltaic, and other industries are discussed, along with the trends of wafer technology and their benefits. Various properties of wafers are presented including the surface properties, wafer flatness, nanotopography, surface roughness, materials and mechanical properties, crystalline silicon wafer orientation, anisotropy of silicon wafers, and gravity-induced deflection for large wafers. The economics of wafer manufacturing are discussed based on the processes of manufacturing processes.

References

Allan A, Edenfeld D, Joyner WH, Kahng AB, Rodgers M and Zorian Y 2002 2001 technology roadmap for semiconductors. *Computer* **35**(1), 42–53.

American Physical Society 2000 This month in physics history, Nov 17–Dec 23, 1947: Invention of the first transistor. *American Physical Society (APS)*.

Anderson D 1997 Stoking the productivity engine with new materials and larger wafers. *Solid State Technology* **40**, 57.

Applied Materials 2011 Wafer wire sawing economics and total cost of ownership optimization. Technical report, Applied Materials, Inc., 3050 Bowers Ave., Santa Clara, CA 95054-3299.

ASME B46 2009 Surface texture (surface roughness, waviness, and lay). ASME B46.1-2009 (Revision of ASME B46.1-2002).

ASTM Standard F1390 2002 Standard test method for measuring warp on silicon wafers by automated noncontact scanning (withdrawn 2003) ASTM International, West Conshohocken, PA. www.astm.org.

ASTM Standard F1530 2002 Standard test method for measuring flatness, thickness, and thickness variation on silicon wafers by automated noncontact scanning (withdrawn 2003) ASTM International, West Conshohocken, PA. www.astm.org.

ASTM Standard F533 2002 Standard test method for thickness and thickness variation of silicon wafers (withdrawn 2003) ASTM International, West Conshohocken, PA. www.astm.org.

ASTM Standard F534 2002 Standard test method for bow of silicon wafers (withdrawn 2003) ASTM International, West Conshohocken, PA. www.astm.org.

ASTM Standard F657 1999 Standard test method for measuring warp and total thickness variation on silicon wafers by noncontact scanning (withdrawn 2003) ASTM International, West Conshohocken, PA. www.astm.org.

Bhagavat M, Prasad V and Kao I 2000 Elasto-hydrodynamic interaction in the free abrasive wafer slicing using a wiresaw: Modeling and finite element analysis. *Journal of Tribology* **122**(2), 394–404.

Bhagavat S and Kao I 2006 Theoretical analysis on the effects of crystal anisotropy on wiresawing process and application to wafer slicing. *International Journal of Machine Tools and Manufacture* **46**, 531–541.

Bhagavat S, Liberato J, Chung C and Kao I 2010 Effects of mixed abrasive grits in slurries on free abrasive machining (FAM) processes. **50**, 843–847.

Buijs M and Houten KK 1993a A model for lapping of glass. *Journal of Materials Science* **28**(11), 3014–3020.

Buijs M and Houten KK 1993b Three-body abrasion of brittle materials as studied by lapping. *Wear* **166**(2), 237–245.

Byrne D, Kovak BK and Michaels R 2013 Price and quality dispersion in an offshoring market: Evidence from semiconductor production services. Technical report, Finance and Economics Discussion Series, Division of Research and Statistics and Monetary Affairs, Federal Reserve Board, Washington, D.C.

Chang Y, Hashimura M and Dornfeld D 2000 An investigation of material removal mechanisms in lapping with grain size transition. *Journal of Manufacturing Science and Engineering: Transactions of the ASME* **122**(3), 413–419.

Chauhan R, Ahn Y, Chandrasekar S and Farris T 1993 Role of indentation fracture in free abrasive machining of ceramics. *Wear* **162**, 246–257.

CNSE 2012 The global 450 consortium (g450c) website. URL http://www.g450c.org.

CNSE 2015. SUNY Poly CNSE to present cutting-edge semiconductor technology developments at SEMICON West 2015 Conference. URL https://sunypoly.edu/news/suny-poly-cnse-present-cutting-edge-semiconductor-technology-developments-semicon-west-2015.html.

Edenfeld D, Kahng AB, Rodgers M and Zorian Y 2004 2003 technology roadmap for semiconductors. *Computer* **37**(1), 47–56.

EEMI 2013 The move to the next silicon wafer size *SEMICON, Europa 2013* European 450 mm Equipment & Materials Initiative: EEMI 450.

Gise P and Blanchard R 1986 *Modern Semiconductor Fabrication Technology*. Prentice Hall.

Goloni D and Jacobs S 1991 Physics of loose abrasive microgrinding. *Applied Optics* **30**(19), 2761–2777.

Hopcroft MA, Nix WD and Kenny TW 2010 What is the young's modulus of silicon?. *J. of Microelectromechanical Systems* **19**(2), 229–238.

Intel Corp. 2005 Moore's law URL http://download.intel.com/museum/Moores_Law/Printed_Materials/Moores_Law_2pg.pdf.

Intel News Release 2008 Intel, Samsung Electronics, TSMC reach agreement for 450 mm wafer manufacturing transition Website. URL http://www.intel.com/pressroom/archive/releases/2008/20080505corp.htm.

ISO-4287 1996 Iso 4287-1: Surface roughness - terminology - part 1: Surface and its parameters.

Kao I 2004 Technology and research of slurry wiresaw manufacturing systems in wafer slicing with free abrasive machining. *the International Journal of Advanced Manufacturing Systems* **7**(2), 7–20.

Leachman RC, Ding S and Chien CF 2007 Economic efficiency analysis of wafer fabrication. *IEEE Trans on Automation Science and Engineering* **4**(4), 501–512.

Li J, Kao I and Prasad V 1998 Modeling stresses of contacts in wiresaw slicing of polycrystalline and crystalline ingots: Application to silicon wafer production. *Journal of Electronic Packaging* **120**(2), 123–128.

Mitutoyo 2016 Mitutoyo america corporation bulletin no. 2229: Quick guide to surface roughness measurement, reference guide for laboratory and workshop URL. https://www.mitutoyo.com/wp-content/uploads/2012/11/1984_Surf_Roughness_PG.pdf.

Moore GE 1965 Cramming more components onto integrated circuits. *Electronics Magazine* **38**(8), 4–7.

Precision Device 2016 Surface roughness terminology and parameters URL. https://www.predev.com/pdffiles/surface:roughness_terminology_and_parameters.pdf.

Preston FW 1927 The structure of abraded glass surfaces. *Trans. Opt. Soc.* **23**, 141–164.

Riordan M and Hoddeson L 1997 *Crystal Fire: The Birth of the Information Age*. W. W. Norton & Company.

SEMI *DRAFT Document 3089: Guide for Reporting Wafer Nanotopography*. SEMI.

SEMI 2012a 450 mm central news and information website. URL http://www.semi.org/en/Issues/450 mm.

SEMI 2012b Background on 450 mm wafer transition website. URL http://www.semi.org/en/node/42896.

SEMI 2014 SEMI reports shift in semiconductor capacity and equipment spending trends website. URL http://www.semi.org/en/node/48566.

SEMI 3D12 2015 Guide for measuring flatness and shape of low stiffness wafers (3D12-0315) website. URL http://www.semi.org, https://www.semiviews.org.

SEMI M40 2014 Guide for measurement of roughness of planar surfaces on polished wafers (M40-1114) website. URL http://www.semi.org, https://www.semiviews.org.

SEMI M43 2018 Guide for reporting wafer nanotopography (M43-0418) website. URL http://www.semi.org, https://www.semiviews.org.

SEMI M78 2018 Guide for determining nanotopography of unpatterned silicon wafers for the 130 nm to 22 nm generations in high volume manufacturing (M78-0618) website. URL http://www.semi.org, https://www.semiviews.org.

SEMI MF1390 2018 Test method for measuring bow and warp on silicon wafers by automated noncontact scanning (MF1390-0218) website. URL http://www.semi.org, https://www.semiviews.org.

SEMI MF1530 2018 Test method for measuring flatness, thickness, and total thickness variation on silicon wafers by automated noncontact scanning (MF1530-0707) website. URL http://www.semi.org, https://www.semiviews.org.

SEMI MF1811 2016 Guide for estimating the power spectral density function and related finish parameters from surface profile data (MF1811-1116) website. URL http://www.semi.org, https://www.semiviews.org.

SEMI MF533 2010 Test method for thickness, and thickness variation of silicon wafers (MF533-0310) website. URL http://www.semi.org, https://www.semiviews.org.

SEMI MF534 2014 Test method for bow of silicon wafers (MF534-0707) website. URL http://www.semi.org, https://www.semiviews.org.

SEMI MF657 2014 Test method for measuring warp and total thickness variation on silicon wafers by noncontact scanning (MF657-0707) website. URL http://www.semi.org, https://www.semiviews.org.

Watanabe M and Kramer S 2006 450 mm silicon: An opportunity and wafer scaling. *The Electrochemical Society Interface* pp. 28–31.

Wei S and Kao I 1999 Hight-resolution wafer surface topology measurement using phase-shifting shadow moiré technique In *the Proceedings of IMECE'99: DE-Vol 104, Electronics Manufacturing Issues* (ed. Sahay C, Sammakia B, Kao I and Baldwin D), pp. 15–20. ASME Press, Three Park Ave., New York, NY 10016.

Wei S, Wu S, Kao I and Chiang FP 1998 Wafer surface measurements using shadow moiré with Talbot effect. *Journal of Electronic Packaging* **120**(2), 166–170.

Wortman JJ and Evans RA 1965 Young's modulus, shear modulus, and poisson's ratio in silicon and germanium. *J. Appl. Phys.* **36**(1), 153–156.

Xie Y and Bhushan B 1996 Effects of particle size, polishing pad and contact pressure in free abrasive polishing. *Wear* **200**(1-2), 281–295.

Yang F and Kao I 1999 Interior stress for axisymmetric abrasive indentation in the free abrasive machining process: Slicing silicon wafers with modern wiresaw. *Journal of Electronic Packaging* **121**(3), 191–195.

Yang F and Kao I 2001 Free abrasive machining in slicing brittle materials with wiresaw. *Journal of Electronic Packaging* **123**, 254–259.

Young R 2011a 450 mm industry briefing *SEMICON Japan*. Opening Remarks.

Young R 2011b The 450 mm transition: Consortium status, plans, and strategy website. URL http://www.sematech.org/meetings/archives/symposia/10202/Session%205-450/450mm_01_Young.pdf.

Zhu L and Kao I 2005 Galerkin-based modal analysis on the vibration of wire-slurry system in wafer slicing using wiresaw. *Journal of Sound and Vibration* **283**(3-5), 589–620.

2

Wafer Manufacturing: Generalized Processes and Flow

2.1 Introduction

Wafer manufacturing, or wafer production, refers to a series of processes that make wafers of different shapes and materials for microelectronics fabrication and other applications. Prime wafers are produced from various materials, such as silicon, silicon carbide, lithium niobate, gallium arsenide, III–V semiconductor materials, and optoelectronics materials. Wafer manufacturing processes are also employed to produce photovoltaic (PV) wafers from single or polycrystalline silicon. Figure 2.1 illustrates the process flow and categories of wafer manufacturing. A wafer is a thin slice of material, usually in a round shape, that is prepared with mirror-finish surface ready for device fabrication. Wafer manufacturing has been brought to the limelight largely due to semiconductor and microelectronics fabrication in the second half of the twentieth century. With the advent of silicon-based solar wafer technology in the 1990s, a wave of new machine tools were deployed to produce silicon wafers. In the few years that followed, the microelectronics industry started to adopt wiresaws as the mainstream technology for wafer slicing, resulting in a wholesale change of manufacturing lines from inner-diameter (ID) saws to wiresaws. These machine tools facilitate the development of larger wafers of diameters 300 mm, or larger, in microelectronics fabrication.

2.2 Wafer Manufacturing: Generalized Process Flow

The generalized process flow of wafer manufacturing is illustrated in Figure 2.1, including the following classification of four categories [Kao (2004)]:

- Crystal growth: the end product is a crystal ingot or boule with prescribed crystalline characteristics of the source materials.
- Wafer forming: the end product is a sliced wafer of the prescribed dimension and thickness.

Wafer Manufacturing: Shaping of Single Crystal Silicon Wafers,
First Edition. Imin Kao and Chunhui Chung.
© 2021 John Wiley & Sons Ltd. Published 2021 by John Wiley & Sons Ltd.

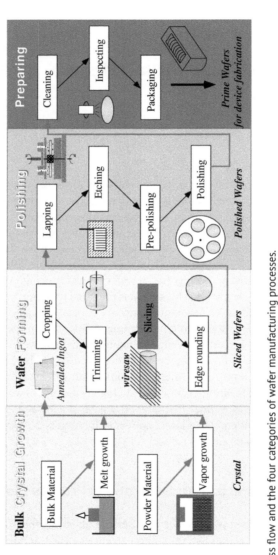

Figure 2.1 Process flow and the four categories of wafer manufacturing processes.

- Wafer polishing: the end product is a wafer with a polished surface that is ready for microelectronics fabrication; some wafers, such as photovoltaic wafers, may not require this process.
- Wafer preparing: the end product is a package of prime wafers ready for micro-electronics device fabrication in a clean room environment.

In the following sections, each of the four categories will be discussed.

2.3 Crystal Growth

Growing crystals is the first step toward wafer manufacturing. Although there are many technologies for growing crystal, the two most common approaches are *melt growth* and *vapor growth*, as illustrated in Figure 2.1. Melt growth produces bulk crystal ingots or boules of different diameters, which can exceed 300 mm, with meter-long lengths. Vapor growth typically produces a thin layer of polycrystalline crystal that resembles a sliced wafer. Single crystalline ingots are mostly produced through the melt growth processes, which are the most common technique to produce crystal ingots. For example, the traditional silicon-based semiconductor fabrication uses silicon material grown in different crystalline orientations, utilizing melt growth, which will be described in the following sections.

2.3.1 Melt Growth

Presently, two prevalent methods are employed to grow single crystalline ingots for semiconductor applications. They are the Czochralski (CZ) crystal growth and the float-zone (FZ) crystal growth techniques. The CZ method is the most popular method that accounts for more than 80% of all the silicon crystals grown for the semiconductor industry. With the trend of wafers with larger diameter, the CZ method is the most suitable technique for crystal growth.

The Czochralski Growth (CZ)
The Czochralski growth (CZ) process is named after Polish scientist Jan Czochralski who invented this method in 1916 while studying the crystallization of metals. The CZ ingot growth for silicon ingots starts with pieces of polycrystalline silicon of high purity, called the "charge," as those shown in Figure 1.2. These charges are melted in a quartz crucible (SiO_2), as illustrated in Figure 2.2, along with small quantities of specific *impurity* elements in group III and group V of the periodic table, called dopants[1]. The materials in the crucible are then heated to a temperature above the melting point, about 1420 °C for silicon. The quartz crucible, shown in Figure 2.2, is

---

1 If the dopant element is in group III, the resulting ingot will be P type; whereas, if the dopant is in group V, the ingot will be N type. The most common dopants for P types are boron, aluminum and gallium, and for N types phosphorus, arsenic and antimony. The dopants determine if the crystal is P or N type which dictates the electrical properties of the wafers and the microelectronics fabrication processes.

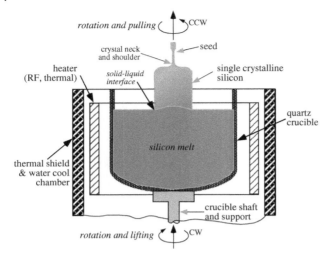

Figure 2.2 An illustration of the Czochralski (CZ) crystal growth

heated by induction using radio frequency (RF) heating or thermal resistance heating. Once the polycrystalline silicon and dopant material have been melted, a *seed* crystal made of a single crystalline silicon crystal, with a prescribed crystalline orientation, is dipped into the molten silicon, called the "melt," and then withdrawn at a properly controlled rate. Under carefully controlled process parameters of rotation and pulling, as illustrated in Figure 2.2, the material in the melt will solidify (or freeze) at the solid–liquid interface on the surface of the melt that accurately replicates the crystalline structure of the seed crystal. The seed is held by a seed holder (not shown in Figure 2.2), which pulls and rotates the crystal solid. The pulling speed is the fastest when growing the narrow neck (about 120–300 mm h^{-1}) to minimize the initial dislocation defects. Once the neck is formed, the pulling speed is reduced in order to properly control and form the shoulder to transition to the desired size of the ingot with a steady-state growth thereafter. A photo of a single crystalline silicon ingot grown by the CZ method is shown in Figure 2.3. The figure illustrates, from the top, the seed section, followed by the neck section and the shoulder to transition to the bulk section of the ingot with target diameter of 200 mm. As the solid is slowly raised above the melt, the surface tension between the seed and the melt causes a thin film of the silicon to adhere to the seed and then to cool. While cooling, the atoms in the melted silicon orient themselves to the crystalline structure of the seed. As a result, the finished ingot will possess the same crystalline orientation as the seed. The total time required to grow a silicon ingot can be a week to a month, depending on many factors, notably the size and quality [Gise and Blanchard (1986)].

To achieve uniformity and to avoid the formation of local hot or cold zones, the crucible with melt rotates in one direction at 12–14 rpm while the seed holder rotates in the opposite direction at half of that angular speed at 6–8 rpm. During the growth process, a crucible shaft and support mechanism, as illustrated in Figure 2.2, also

Figure 2.3 A single crystalline silicon ingot produced by CZ bulk growth, showing the section of seed with neck and shoulder. Source: Imin Kao.

lifts the crucible system to maintain a consistent solid–liquid interface since the melt level in the crucible will drop relative to the hot-zone area of the furnace, as the crystal ingot is grown and pulled. The growth process is carefully monitored and controlled to produce ingots of the desired diameter. Insert gas, such as argon, is often used as the ambient gas in the growth chamber.

In modeling CZ crystal growth, the relationship between the pulling speed and crystal diameter can be derived by formulating and solving differential equations [Plummer et al. (2000)]. The following equation of the maximum pulling rate can be obtained:

$$V_{\max} = \frac{1}{LN} \sqrt{\frac{2\sigma\epsilon k_M T_M^5}{3r}} \tag{2.1}$$

where $V_{\max}$ is the maximum pulling speed of the crystal ingot, L is the latent heat of fusion, N is the density of the crystal material, $\sigma = 5.670373 \times 10^{-8}$ J s^{-1} m^{-2} K^{-4} is the Stefan–Boltzman constant, ϵ is the emissivity of the crystal material, k_M is the thermal conductivity at the melting point, T_M is the melting temperature, and r is the radius of the crystal ingot. Equation (2.1) suggests that the pulling speed is inversely proportional to the square root of the radius of the ingot. An illustration of the theoretical pulling speed of silicon in CZ growth, as a function of the diameter of ingots, is shown in Figure 2.4. The two dots in the plot indicate the pulling

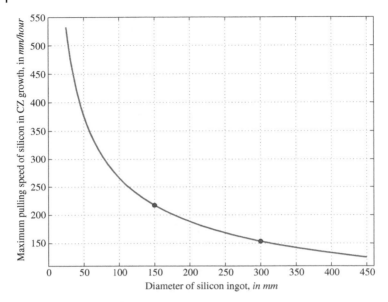

Figure 2.4 A plot of maximum theoretical pulling speeds of silicon in CZ bulk growth as a function of the diameters of ingots.

speeds corresponding to the ingot diameters of 150 mm (6 in) and 300 mm (12 in), respectively.

In the case of the growth of silicon ingots, the following values in SI units can be used: $L = 1.7877 \times 10^6$ J/kg, $N = 2329$ kg m^{-3} at room temperature and 2570 kg m^{-3} at the melting point, $\epsilon = 0.55$, $k_M = 20$ J s^{-1} m^{-1} K^{-1} at the melting point[2], and $T_M = 1687$ K or 1414 °C. As an example, the maximum pulling rate of a silicon ingot of a diameter of $r = 150$m (6 in) can be calculated using Equation (2.1), as follows:

$$V_{max} = \frac{1}{(1.7877 \times 10^6)(2570)} \sqrt{\frac{2(5.670373 \times 10^{-8})(0.55)(20)(1687)^5}{3(0.150/2)}}$$
$$= 6.0 \times 10^{-5} \text{ m s}^{-1} = 215 \text{ mm h}^{-1}. \tag{2.2}$$

Thus, the maximum pulling rate for the CZ growth of a silicon ingot of a diameter of 150 mm, or 6 in, is about 215 mm h^{-1}. The actual pulling rates are smaller than this theoretical maximum rate. Referring to Equation (2.1) and Figure 2.4, the maximum theoretical pulling speed of a 450 mm silicon ingot in CZ growth is 125 mm h^{-1}.

The Liquid Encapsulated Czochralski Growth (LEC)
Liquid encapsulated Czochralski, also known as the LEC technique, makes it possible to grow single crystals from materials that contain chemical elements that produce high vapor pressure at the melting point. This modified method of the

---

2 The thermal conductivity of silicon is dependent upon the temperature. At room temperature, the thermal conductivity is about 150 J s^{-1} m^{-1} K^{-1}. At the melting point, T_M, the thermal conductivity is about 20 J s^{-1} m^{-1} K^{-1} [Glassbrenner and Slack (1964)].

Czochralski technique is widely adopted to grow single crystal of III–V compound semiconductors, such as GaAs. In the growth of GaAs using the LEC method, boron trioxide, with lower melting point than the crystal materials that are either pre-synthesized polycrystalline chunks or the elemental Ga and As, is used to encapsulate the entire melt in the crucible [Hurle (1994); Mullin (1993); Mullin et al. (1965)]. These source materials of the crystal are placed in the CZ growth crucible along with a pellet of boron trioxide. At approximately 460 °C, the boron trioxide melts to form a layer of thick and viscous liquid which coats the entire melt of crystal to be grown in the crucible. This layer of boron trioxide, together with the pressure of inert gas in the CZ crystal growth chamber, prevents sublimation of the volatile group V element. The temperature is increased until the compound synthesizes. A seed crystal is then dipped, through the boron trioxide layer, into the melt. The seed is rotated and slowly pulled as a single crystal crystal ingot is formed.

According to Jurisch and Eichler (2003), the GaAs single crystal was first grown by the LEC method in 1964 by RSRE. The diameter of the GaAs crystal has since increased to 2 in, grown by the Bell Lab in 1971, 100 mm by Westinghouse in 1984, 150 mm by Litton Airtron in 1991, and 200 mm by FCM in 2000.

Much research has been performed on CZ crystal growth; some are listed here for further reference: Dhanaraj et al. (2010); Feigelson (2004); Fickett and Mihalik (2001); Hurle (1994); Krause et al. (2003); Lu and Kimbel (2011); Mullin (1993); Noghabi et al. (2011); Wang et al. (2003, 2006).

Float-zone Growth (FZ)

Float-zone growth (FZ) utilizes a uniform and crack-free cylindrical polysilicon rod in the silicon purification process. As illustrated in Figure 2.5, the polycrystalline silicon rod is configured at the top of the growth chamber which is filled with inert gas. A seed crystal with the desired crystalline orientation is placed so that it is in contact with the bottom side of the polycrystalline silicon rod. An induction heating coil is placed around the rod to melt a small length or zone of the rod, starting with part of the seed. The molten region at the interface between the solid seed and the melt of the rod solidifies and assumes the same crystalline orientation as that of the seed. The float zone is about 20 mm. This floating zone of melt gradually moves upwards as the solidification of the single crystalline ingot forms in the lower part of the growth chamber, as shown in Figure 2.5. The diameter of the ingot is determined by the motion of the heating coil as it travels upwards. The speed at which the heating coil moves is properly controlled to prevent the floating zone from collapsing in the process. The FZ growth method is suitable for an ingot of small diameter, typically not larger than 75 mm.

The FZ growth method does not require a crucible (SiO_2), and produces ingots of low oxygen and carbon contents with higher resistivity than those of the CZ growth. With higher purity and fewer micro defects, FZ growth can produce 10–20% higher efficiency in solar cells. In addition, the mechanism of growth allows for faster growth rates and heating–cooling time at localized zone and provides a substantial economic advantage. Silicon wafers of extremely high purity can be made using FZ growth. Such crystal wafers are preferred in device fabrication such as power

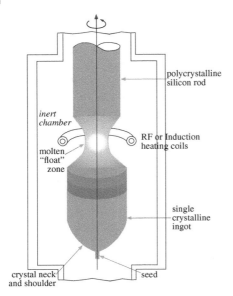

Figure 2.5 Float-zone (FZ) growth.

diodes, power transistors, thyristors, rectifiers, and silicon-controlled rectifiers (SCRs).

Comparison Between the CZ and FZ Growth Methods

Table 2.1 compares the characteristics between the two methods of crystal growth for single crystalline ingots. It is noted that the CZ method has a higher speed of growth than FZ. In addition, the CZ growth method can produce ingots of larger diameters – a very important advantage in today's silicon industry requiring ingots of large diameters. However, FZ growth can produce ingots of lower oxygen and carbon content, desirable for certain devices. The FZ technique can also be used for the purpose of purification, called zone refining. Growing ingots using the CZ method requires a crucible and more moving parts; hence, it demands higher initial cost and maintenance expenses.

Table 2.1 Comparison between the two melt growth methods: CZ and FZ growth.

Characteristics	CZ growth	FZ growth
Growth speed (mm/hour)	60–120	180–300
Requires crucible	Yes	No
Diameter of ingot	Large (up to 450 mm)	Small (up to 75 mm)
Electrical power	Higher (45 to 60 kW h kg^{-1})	Lower (25 to 30 kW h kg^{-1})
Dislocation	None	None
Oxygen content (atoms/cm^3)	Higher	Lower $< 1 \times 10^{16}$
Carbon content (atoms/cm^3)	Higher	Lower $< 1 \times 10^{16}$

Bridgman Growth Method

The Bridgman growth method, also known as Bridgman–Stockbarger growth, is named after Percy W. Bridgman and Donald C. Stockbarger, for growing both single and polycrystalline ingots using temperature-controlled cooling in a typically two-zone furnace. The Bridgman method is popular in producing III–V semiconductor compounds, II–VI compounds, and certain piezoelectric crystals. It can achieve a low level of thermal stress and low dislocation density when the process is well controlled [Cröll and Volz (2009)].

This method of crystal growth utilizes a temperature gradient as the furnace moves from hot to cool relative to the melt contained in an ampoule. The configuration of the growth furnace can be in a horizontal (as in Figure 2.6) or a vertical (as in Figure 2.7) arrangement. The polycrystalline material is often sealed in high vacuum in a quartz ampoule and heated above its melting point in the hot zone of furnace with a controlled temperature distribution. A single crystal seed of the desired crystallographic orientation is used to progressively propagate through

Figure 2.6 Schematic of Bridgman growth, in a horizontal configuration. The temperature T_m is the melting point of the crystalline material.

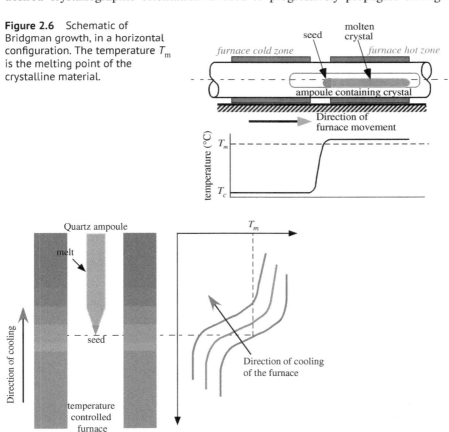

Figure 2.7 Schematic of Bridgman–Stockbarger growth in a vertical configuration with a temperature-controlled furnace. The temperature T_m denotes the melting point of the crystalline material.

the melt to form the single crystalline ingot as the melt slowly solidifies to form a crystalline ingot.

The difference between the Bridgman and Stockbarger methods lies in the control of temperature gradient as the melt solidifies. The Bridgman method uses a two-zone uncontrolled temperature gradient, as illustrated in Figure 2.6. In this configuration, the two-zone furnace moves relative to the ampoule from left to right slowly, producing gradual cooling to the melt in the ampoule. The temperature distribution illustrates that the hot zone is above the melting point (T_m), while the cool zone is at a lower temperature (T_c). On the other hand, the Stockbarger method utilizes a better control of temperature gradient at the melt–solid interface with a temperature-controlled furnace. As illustrated in Figure 2.7, the temperature-controlled furnace gradually lower the temperature below the melting point, T_m, as it cools the melt in the quartz ampoule from bottom to top, which solidifies the melt to produce crystalline ingot. The temperature profiles across the length of the furnace show the effect of gradual cooling of the melt. Figure 2.7 can also be configured such that the furnace moves upwards to gradually cool the melt in the ampoule to produce single crystalline ingot. This process usually takes long time, for example, 10–15 days, depending on the size of the crystal.

However, the name *Bridgman growth* method today generally has been used to refer to an array of growth techniques, configured in horizontal and vertical arrangement, such as those in Figures 2.6 and 2.7, with, or without, temperature gradient control.

Furthermore, the Bridgman growth technique is rarely employed for the production of silicon ingots. Nevertheless, it is frequently applied to grow single crystalline gallium arsenide – a III–V semiconductor compound. When utilizing the horizontal two-zone Bridgman growth furnace for the growth of gallium arsenide, as illustrated in Figure 2.6, the hot zone is kept at a temperature slightly above the melting point at $T_m = 1240\,°C$, while the cool zone is kept at $T_c = 610\,°C$. The polycrystalline GaAs raw material stock in the ampoule is held in the hot zone to melt, while allowing sufficient additional arsenic within the ampoule to prevent the loss of arsenic from the GaAs melt.

Other Melt Growth Methods

There are other melt growth techniques, for example, the Kyropoulos technique. As in the CZ growth method, the seed is brought into contact with the melt but the crystal is grown with a flat top before the initiation of pulling [Dhanaraj et al. (2010)]. This tends to reduce the incidence of twins. At this time, the pulling will start when the crystal has reached the intended diameter during the growth along the direction perpendicular to the growth axis of the crystal. A magnetic field is needed to stabilize the growth of the flat top before pulling. The Kyropoulos method is often found in the growth of alkali halides, to avoid the cracking problems of container growth, that are used to make optical components.

It is interesting to note that only three crystal techniques were recognized in a review paper by Wells (1946). They were the Bridgman method, the Kyropoulos method, and solid-state recrystallization [Dhanaraj et al. (2010)]. The CZ growth method was not considered as a technique to grow large crystals then. After the invention of the first transistor in 1947 by the Bell Lab (cf. Section 1.2.2), the growth of germanium and silicon was demonstrated by the Bell Lab and Texas Instruments, with the name of the crystal growth technique assigned to Czochralski [Dhanaraj et al. (2010)].

2.3.2 Vapor Growth

Solid crystals can be grown by means of vapor or gaseous molecules attaching themselves onto a solid surface under appropriate conditions and configuration. This is commonly referred to as "vapor growth." Such vapor growth process requires a greater rate of deposition of the vapor or gas molecules onto a solid surface than the rate of the molecules on the solid surface to become gaseous. This non-equilibrium state is called supersaturation. Supersaturation is achieved by maintaining the solid crystal at a lower temperature than the gas. A crystal seed with a prescribed crystalline structure and orientation is employed for the gas molecules to replicate the same crystalline structure while attaching themselves to the surface of the seed crystal. Vapor growth is a slow process that proceeds one molecule at a time and builds one layer at a time. Although the process is slow, the vapor growth method can produce very pure crystals. The process is maintained at temperatures well below the melting point of the crystal in order to reduce the density and defects.

Growth of crystal from vapor can be described by the vapor–liquid–solid (VLS) mechanism [Wagner and Ellis (1964)]. Such methodology has widespread applications in the growth of filaments, bulk crystals, and epitaxial layers.

2.3.3 Epitaxial Growth

Epitaxial growth is a vapor growth process, which refers to a growth process to produce a desired layer of crystalline deposition on a crystalline substrate. The overlaid layer is called an epitaxial layer or epitaxial film or epi[3] layer. For applications in semiconductor fabrication, the deposited material should ideally form a single crystalline layer that has a well-defined crystalline orientation with respect to the substrate crystalline structure, as illustrated in Figure 2.8. The interface is established when the process of epitaxial growth starts. This is often referred to as the "homoepitaxy" in which a crystalline layer is grown on a substrate of the same material.

In order for the epitaxial growth process to produce the desired outcome, the wafer substrate is ground or lapped, cleaned, and then polished, followed by dipping in a series of chemical agents, such as sulfuric acid, hydrogen peroxide, then rinsing with water and deionized water and dried with isopropyl alcohol. Figure 2.9 shows such a process. This etching cycle is repeated, followed by each crystal layer growth cycle.

---

3 Epitaxial growth process or its outcomes related to the epitaxial process is often abbreviated as "epi," such as epi-growth or epi-layer. Similarly, homoepitaxial is often referred to as "homoepi."

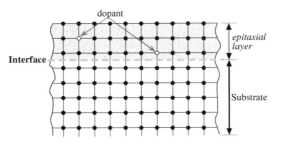

Figure 2.8 Epitaxial growth on a wafer substrate depicting the interface at the epitaxial layer with the substrate, as well as dopant distribution. The density of the dopant in the epitaxial layer is typically smaller than that of the substrate.

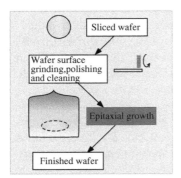

Figure 2.9 Illustration of epitaxial growth on a wafer.

The epitaxial process is used in silicon-based semiconductor manufacturing processes for logic gates, bipolar junction transistors, complementary metal-oxide-semiconductor (CMOS), III–V-based optical devices, and others. Issues of consideration in epitaxial processes include the uniformity of the deposition, resistivity, dopant density, and thickness, as well as process control to maintain the purity, to reduce the defects, and to protect the surfaces during the manufacturing process and the handling afterwards.

Silicon wafers made of the epitaxial process are usually grown using vapor-phase epitaxy, a modification of the chemical vapor deposition (CVD) process. Molecular beam and liquid phase epitaxy are also used, primarily for III–V and II–IV and other compound semiconductor materials. Epitaxial silicon is often grown by two source gases, silicon tetrachloride and hydrogen, at approximately 1200 °C

$$SiCl_4 + 2\ H_2 \leftrightarrow Si + 4\ HCl. \tag{2.3}$$

The reaction has two sources gases ($SiCl_4$ and H_2) which produce a solid (Si) for the epitaxial deposition and another gas (HCl). Polycrystalline silicon will be grown if the growth rate is above two microns per minute. On the other hand, if too much hydrogen chloride (HCl) byproduct is present, the etching reaction of the wafer will compete with the deposition of silicon, and can result in a negative growth rate when the etching outpaces the deposition. Thus, the control of the process and parameters are very important to produce a single crystalline silicon epi-layer on top of the silicon substrate with consistent orientation alignment. The vapor phase epitaxy process of silicon can also use silane, dichlorosilane, or trichlorosilane as the source gas. For example, the silane reaction can also produce silicon for deposition as follows

$$SiH_4 \rightarrow Si + 2 H_2 \tag{2.4}$$

with reaction taking place at a lower temperature of 650 °C. The reaction does not produce byproduct (HCl) which will etch the silicon; however, the reaction produces a polycrystalline silicon epi-layer, and allows oxidization species to contaminate the epi-layer, unless the epi-growth process is meticulously controlled.

2.3.4 Casting Polycrystalline Crystal

Photovoltaic wafers are often sliced from polycrystalline silicon crystal by melting the silicon raw material and cast for cooling. The resulting polycrystalline silicon ingot often has a square cross-sectional area. Preventing cracks from forming and maintaining the homogeneity of the cast during the thermal cooling process is key to the success of producing such crystals.

2.3.5 Other Crystal Growth Methods

In addition to the two most common approaches of crystal growth described in this section (Section 2.3) and the various methods associated with the two approaches presented in the preceding sections, other methods of crystal growth have been developed and used to growth crystals of different materials.

Pamplin (1980) presented four approaches of crystal growth as (i) melt growth (Section 2.3.1), (ii) vapor growth (Section 2.3.2), (iii) solid-state growth, and (iv) solution growth. This is illustrated in Figure 2.10, with certain individual methods illustrated in the preceding sections. The flux growth method, for example, is a solution growth that requires relatively simple equipment. It can grow congruently and incongruently melting materials. Although the flux method requires less time to grow, it cannot produce crystals of larger diameter [Canfield and Fisk (1992); Fisk and Remeika (1989)].

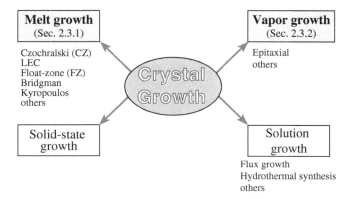

Figure 2.10 Four approaches of crystal growth, along with reference to the section numbers and methods. Source: Based on Pamplin BR (1980).

2.4 Wafer Forming

The wafer manufacturing process following crystal growth is the *wafer forming* process which includes cropping, trimming, orientation identification, slicing and edge rounding, which are presented in the following sections.

2.4.1 Cropping

Once an ingot is grown, it is cropped at both ends. The seed can be re-used. The tail end of the growth process typically contains a higher concentration of impurities and needs to be removed. When both ends are properly removed, the end result is a cropped crystal ingot.

2.4.2 Trimming

The cropped crystal ingot is then trimmed to a rough-size diameter a little larger than the desired diameter of the finished silicon wafer. A diamond grinder is often used for this trimming process to shape the form of the ingot to a uniform cylinder of the specified diameter that is ready for slicing.

2.4.3 Orientation Identification

The ingot of uniform diameter after the trimming process is then ground with the appropriate orientation flat to indicate the crystalline orientation of wafers with a diameter smaller than or equal to 200 mm, as summarized in Table 1.3. For wafers of larger diameters, a notch is used to identify the crystalline orientation of [1 1 0], as illustrated in Figure 1.3.

The seed dictates the crystal face on the wafer surface; however, the other axes of the crystal vary along the perimeter of the wafer and are determined by the rotational position of the crystal ingot. The single crystalline ingot and wafer are anisotropic and have preferential cleavage planes, as illustrated in Section 1.5.2. It is critical for the yield of the fabrication and for the later device separation (scribe and break) to align the dice of circuits precisely with respect to the cleavage planes. This alignment is accomplished through the indication of orientation flats and notch on the wafer. X-ray diffraction is used to determine the crystal orientation before grinding orientation flat(s) or notch.

2.4.4 Slicing

Once having been ground with orientation identification, and having passed a number of inspections, the trimmed ingot is sliced into wafers of slightly larger thickness than the desired specification. This process is sometimes called *wafering*. Wafer slicing is the first step in the *wafer forming* processes of wafer manufacturing. Issues of consideration in slicing process include: the edge flakes, exit damage, bow, surface warp, and subsurface cracks.

Slicing is critical to the success of not only wafer forming but also the subsequent processes of polishing and preparing. Semiconductor materials are typically brittle; hence, many traditional metal machining tools cannot be readily used to slice such brittle materials with the necessary large throughput and quality. While it is necessary to slice the ingot into slices of wafers, the kerf loss due to the slicing process needs to be minimized in order to maximize the number of wafer throughput. Therefore, a different genre of machine tools have been employed over the years for wafer slicing.

The conventional, and the first, technology of machine tool for wafer slicing was the inner-diameter (ID) saws. With the need to slice wafers of smaller thickness (such as PV wafers) or larger diameters (such as 300 mm wafers), modern slurry wiresaws largely replaced the ID saws as the preferred tool of manufacturing in the mid-1990s. (See also Sections 2.4.6 and Chapter 4.) Saws using wires impregnated with abrasive grits such as diamond, boron carbide, or silicon carbide also have been employed in slicing ingots of different materials since the early 2000s (see Chapter 6). In the following sections, each type of machine tool for slicing is introduced with comparison.

2.4.5   Slicing Using the Inner-diameter (ID) Saw

The inner-diameter (ID) saw was the standard machine tool for slicing crystal ingots into wafers, especially for silicon wafers, before the introduction of the wiresaw in the early 1990s. A schematic of an ID saw is illustrated in Figure 2.11. It can be seen in Figure 2.11 that the inside diametral edge of a concentric ring-shaped saw blade performs the cutting of the ingot – hence the name of *inner-diameter saw*. The blade of the ID saw is very thin[4] to reduce the kerf loss, and is stretched radially outward with a very high tension using the clamping and tensioning screws on the

Figure 2.11 Schematic of an inner-diameter (ID) saw for wafer slicing.

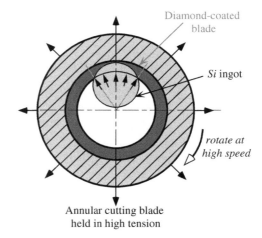

Diamond-coated blade

Si ingot

rotate at high speed

Annular cutting blade held in high tension

---

4 The thickness of the blade varies, depending on the diameter of the inner hole. Typically, the thickness of the blade is 0.1 mm for a bore diameter of about 100 mm, and increases to 0.15 mm for a bore diameter of 200 mm.

ring fixture. The cutting edge of the ID saw is coated with diamond powder to slice through crystal material, such as silicon. The space of the hole in the center needs to accommodate the crystal ingot mounted on a jig. The ID saw spins at high RPM[5], resulting in a blade speed of 18 to 21 m s^{-1} at the cutting or machining interface. The blade speed is adjusted according to the materials being sliced. As the ID blade spins and traverses through the cross section of the ingot, as shown in Figure 2.11, a slice of wafer is obtained when the ingot is completely cut through [Kao (2004)].

2.4.6 Slicing Using a Wiresaw

In this book, the term *wiresaw* is used to refer to a plethora of machine tools that utilize a single wire to span a wire web surface to slice crystal ingots, with abrasive slurry. Often *wiresaw* and *slurry wiresaw* are used interchangeably. Many names can be found in the literature for the manufacturing machine that is called a *wiresaw* in this book. Some called it a "multi-wire saw" which is misleading because only one continuous wire is used to span a wire web that appears to be composed of multiple wires. Some others simply call it a "wire saw" but this can be a reference to so many different types of machine tools using a wire or wires. Therefore, such a machine tool shall be called a "wiresaw" whenever it appears in this book, in order to maintain the specificity of the reference to this specialized machine tool of wire immersed in abrasive slurry for wafering. A schematic of a modern slurry wiresaw is illustrated in Figure 2.12 that has wire supply and take-up spools, providing a thin steel wire to be wound through the grooved surfaces of wire guides to form a "wire web" to cut through the ingot fed on the wire web surface [Bhagavat et al. (2000); Kao (2004); Li et al. (1998)].

As the diameter of silicon wafers increase, the ID saws became less effective. The largest diameter of ingots that can be effectively sliced by ID saws is 200 mm, or 8 in, which is already pushing the limit of the capability of ID saw technology. Wiresaw technology came into the wafer manufacturing industry, first through the door of slicing PV wafers. Due to its high throughput and high quality of sliced surface,

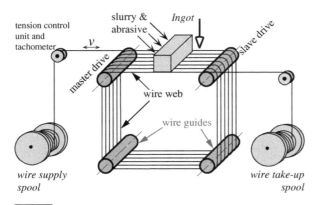

Figure 2.12 Schematic of a modern wiresaw for the slicing of a polycrystalline photovoltaic (PV) silicon ingot into wafers.

tension control unit and tachometer

slurry & abrasive

Ingot

slave drive

v

master drive

wire web

wire guides

wire supply spool

wire take-up spool

---

5 The angular speed of the ID blade ranges from 1500 to 4500 RPM, depending on the the size of bore diameter.

wiresaw technology has become the standard for wafer slicing today. Refer to Chapter 4 for a more detailed treatment on the subject of modern slurry wiresaws.

2.4.7 Other Tools for Slicing

In addition to ID saws and wiresaws, there are other machine tools that are used or being developed to meet the diverse needs of slicing different crystal materials under different requirements and criteria. Here are some of the slicing tools.

Diamond Impregnated or Coated Wire Saw

This type of saw includes a category of many different machine tool designs that utilize a diamond-impregnated wire for slicing. The wire of such type of saw is made of ductile material, with the circumferential surface impregnated or coated by abrasive grits, such as diamond grains, boron carbide, etc. A photo of a diamond-impregnated or diamond-coated saw is shown in Figure 2.13. The saw in Figure 2.13 utilizes a wire with diamond grits impregnated on the surface, as illustrated by the zoom-in view of a segment of the diamond-impregnated wire. The scanning electron microscope (SEM) photo shows that the diameter of the wire is about 200 μm, based on the illustrated scale of 100 μm. Such types of saws can consist of a single wire or multiple parallel wires in contact with the ingot, under different configurations and designs, for slicing. The saw shown in Figure 2.13 uses only one strand of wire, and has only one wire segment pressing on the contact interface with the ingot for slicing.

The main problem of this type of saw is the stripping of abrasive grits on the circumferential surface of the wire. When stripped bare of all its abrasive grits, the wire loses the ability to cut through work material. However, the technology of coating

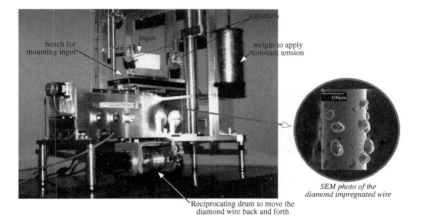

Figure 2.13 Photo of a single-pass, reciprocating saw using a wire impregnated with industrial diamond grits. The saw has a drum with grooves that gathers and feeds a single strand of diamond-impregnated wire, in a reciprocating motion, through the guides and pulleys in order to slice the ingot mounted on the bench using the diamond-impregnated wire. The tension in the wire is maintained through a counterweight, as shown, at about 5 N for slicing. The zoom-in view of a segment of the diamond-impregnated wire is shown by an SEM photo. Source: Imin Kao.

has improved significantly in the last few years to render this technology feasible and competitive to modern wiresaw technology. The presentation of this type of wiresaw can be found in Chapter 6.

Circular Saw

Some saws resemble the circular saw in a conventional carpenter's shop, except the outer edge is coated with diamond abrasives for cutting. Such saws tend to be small in size, due to their vibration and instability compared with ID saws, and are typically used only for customized cuts with small sizes.

2.4.8 Edge Rounding

After the slicing process, the wafers are processed through an edge-rounding machine to remove the square-cornered peripheral edge. This is a necessary step before and after lapping or surface grinding to avoid stress concentration and crack propagation commonly found in brittle materials such as silicon and other semiconductor materials. Due to the nature of the machining process of wafering (see also Chapter 3), micro cracks and burrs will form at the peripheral edge of the wafer and can quickly propagate when stress is applied in handling and processing. This edge grinding procedure drastically reduces the likelihood of cracking or breakage in the remaining steps of wafer manufacturing and later in device fabrication.

After being rounded in a grinding process, the edges need to be polished, depending on the end user's specification. This step improves overall cleanliness and further reduces cracking or breakage up to 400% [SVM (2014)].

2.5 Wafer Polishing

Once an ingot is sliced into wafers of the same thickness, the next group of wafer manufacturing processes serves the purpose of polishing the surface of sliced wafers. There are two main objectives in wafer lapping and polishing processes. They are

1. To polish the surface of wafers by removing surface defects, tool marks of sawing, surface roughness, as well as subsurface damage (SSD).
2. To achieve global planarization of the wafer surface with a very flat surface finish that meets the requirements for fabrication.

The flatness, or global planarization, of the wafer is a critical parameter for the semiconductor industry because the wafer flatness has a direct impact on the subsequent fabrication processes and quality of semiconductor chips diced from the wafer. The wafer polishing processes typically include lapping, etching, and polishing when the diameter of the wafer is 200 mm or smaller. As the wafer diameter increases to 300 mm, the larger surface area and demands for global planarization has prompted the development of surface grinding to improve the efficiency and flatness management of the wafer polishing. In today's industrial applications in wafer manufacturing, the flatness and surface damage removal of silicon wafers are accomplished by a lapping process or grinding process, or a combination of both lapping and grinding. The wafer polishing processes, including lapping, grinding, etching, polishing, chemical-mechanical polishing (CMP), are very important in wafer manufacturing, especially as the size of silicon wafer continues to increase. Wafer surface polishing can be performed on one side that is used for microelectronic fabrication, or on both the front and back surfaces of the wafers.

The individual steps will be briefly discussed, including lapping, grinding, etching, and polishing.

2.5.1 Lapping

After wafers have been sliced and processed, the lapping process begins. Lapping removes tool marks of slicing and surface defects from the front and backside of the wafer. By virtue of applying pressure on the surface of the wafer in the lapping process, it also relieves residual stress accumulated in the wafer from the slicing process. Lapping also removes a layer of surface damage, cracks, pits, and defects, as well as a layer of subsurface damage (SSD) underneath the wafer surface during wafering. Since lapping is also largely a mechanical process, although with less potent abrasive grits [6] compared to those in the slicing process, it creates another shallow layer of SSD. This layer needs to be removed in the next polishing process (Section 2.5.4).

2.5.2 Grinding

Grinding is a process to remove the tool marks on the surface, subsurface cracks, and damage resulting from the wafer forming processes, and to reduce wafer thickness more accurately at a relatively higher material removal rate than lapping. Grinding processes have been used in various steps of the wafer manufacturing. For example, the surface grinding of ingot produces a cylindrical boule of uniform diameter for slicing. The edge grinder removes micro cracks, chips, and burrs from the peripheral edge of wafers by rounding the edge. Nevertheless, using grinding as a new technology to replace lapping in wafer polishing is only a new technological development due to the large 300 mm wafers. This process, when appropriately controlled and managed, can replace lapping, or can be used in conjunction with lapping for the removal of a layer of surface damage, cracks, pits, defects, and at the same time, a layer of SSD typical of an as-sliced wafer.

Proper thickness of wafer (also see Table 1.2) is required in order to handle a 300 mm wafer without breaking during handling and processing. Grinding is also important for semiconductor dice in a back grinding process. Thinning on the back of a diced wafer is useful to reduce electronic thermal resistance [Wood et al. (2013)].

More details regarding lapping and grinding processes will be discussed in Chapter 7 and Section 9.2 in Chapter 9.

---

6 Typical abrasive grits used in the slurry for lapping is aluminum oxide (Al_2O_3), as opposed to the harder abrasives for slicing, such as silicon carbide, boron carbide, or diamond grits.

2.5.3 Etching

Once the silicon wafers are lapped or surface-ground, they go through an etching and cleaning process using sodium hydroxide (NaOH) or acetic acid (CH_3COOH), nitric acids (NH_3), and hydrofluoric acid (HCL) solution to alleviate microscopic surface cracks and nanotopological surface damage caused by lapping or surface grinding.

2.5.4 Polishing

The final and most critical step in wafer manufacturing is polishing which produces a mirror surface finish. This process takes place in a clean room[7] [PCA (2014); SVM (2014)]. Most prime grade silicon wafers go through two to three stages of polishing using progressively finer slurries or polishing compounds. Wafers are typically polished on one side only, except the 300 mm wafers which are polished on both sides in order to achieve a very high degree of flatness. The polished side is used for device fabrication. This surface must be free of topography, micro cracks, subsurface damage, scratches, and residual work damage.

There are two main steps in the polishing process, although the actual industrial practice may vary depending on preference and process design. The first step is stock removal using polish pad and polishing slurry with abrasives, which remove a very thin layer of silicon and produces a wafer surface that is nearly damage free. Such polishing equipment basically comprises a polishing head that holds the semiconductor wafer and polishing pads held by a polishing platen to perform polishing with a slurry consisting of very fine abrasive particles and chemical liquid. Chemicals in the slurry react with and remove the damaged material on the wafer surface.

The second step is CMP using a solution containing sodium hydroxide (NaOH), colloidal silica, and water. This final CMP process does not remove material. Its purpose is to remove a haze from the stock removal process to produce wafers with mirror surface finish for device fabrication. The details of CMP process will be discussed in Chapter 8.

2.6 Wafer Preparation

2.6.1 Cleaning

The main purpose of cleaning is to remove surface contaminants, metal residue and silicon oxide layer. According to the RCA cleaning steps, the wafer is soaked in deionzied water, followed by alkaline-peroxide solution to remove contaminants.

---

7 Clean rooms are rated and range from class 1 to class 10,000. The rating corresponds to the number of particles per cubic foot. For reference, these particles are not visible to the naked eye and in an uncontrolled atmosphere, such as a living room or office, the particle count would likely be five million per cubic foot. To help maintain the level of cleanliness required for microelectronics fabrication, the workers must wear clean-room gowns that cover their body from head to toe that are designed not to collect or carry any particles. They must also stand in a vacuum that blows away any small particles that might have accumulated before entering the room.

The SC-1 solution is composed of deionized water, NH_4OH (ammonium hydroxide, 29% by weight of NH_3) and H_2O_2 (hydrogen peroxide, 30%), in a ratio of about 5:1:1 at 75–80 °C for 10 min. This treatment will result in a layer of silicon dioxide formation and metallic ion contamination on the wafer surface. These oxides and contaminants are removed through short immersion in a 1:100 or 1:50 solution of hydrogen fluoride (HF) and H_2O at 25 °C for about 15 s. This step is optional if ultra pure material is not available. The effective treatment to remove silicon dioxide and metallic ion is performed by immersion of wafer in the SC-2 solution consisting of deionized water, HCl (hydrochloric acid, 39% by weight) and H_2O_2 (hydrogen peroxide, 30%) in a ratio of about 5:1:1 at at 75–80 °C, for 10 min. These standard steps of wafer cleaning are followed by high temperature processing.

The cleaning process is very critical before wafers are inspected and packaged. These procedures can be repeated in as many cycles as needed in order to achieve highly pure wafer. It remains a challenge and opportunity for wafer manufacturing to improve the cleaning process for quality assurance, efficiency, environmental friendliness, and cost-effectiveness. The cleaning method can be modified by different provider in order to optimize quality of the wafer.

More detailed treatment on this subject and the RCA clean method will be presented in Section 9.6.

2.6.2 Inspection

After the process of polishing and final cleaning, wafers are sorted to specification and inspected to ensure the surface quality and readiness for semiconductor fabrication. Defects and unwanted particles can be detected using optical systems, such as a high intensity light and laser scanning system.

2.6.3 Packaging

All wafers are packaged in cassettes and sealed with tapes if they meet the specifications of the inspection standard. To prevent air particles and moisture from entering the cassettes, they are placed in a vacuum sealed plastic bag before leaving the clean room.

2.7 Industrial Processes of Wafer Manufacturing

In this section, we present industrial practice of wafer manufacturing with illustrations of wafer manufacturing equipment and processes, based on the preceding sections of the wafer manufacturing processes, as outlined in Figure 2.1, with the following four categories:

- Crystal growth
- Wafer forming
- Wafer polishing

● Wafer preparing.

Wafer manufacturing processes are illustrated with photos taken in a factory setting.

2.7.1 Crystal Growth

Czochralski (CZ) crystal growth equipment with a view window is illustrated in Figure 2.14. In Figure 2.14(left), a view window attached to the growth equipment is used to monitor the CZ crystal growth process. The photo in Figure 2.14(right), taken from the view window, shows the growth process and the solid–melt interface during the CZ growth process. A CCD camera is connected to the growth equipment directly and an image of the growth interface and process can be acquired anytime during the process. A real-time CCD camera image is illustrated in Figure 2.15.

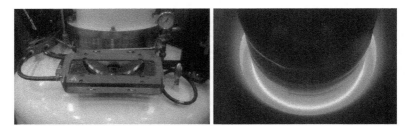

Figure 2.14   (a) Left: CZ crystal growth equipment and a view window. Right: Image of crystal growth viewed from the view window during the growth process. Source: Reproduced by permission of GlobalWafers Co., Ltd.

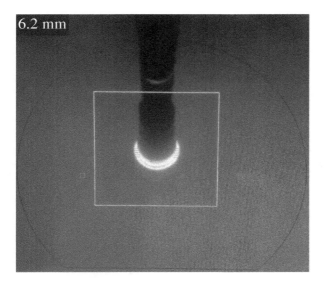

Figure 2.15   A real-time CCD camera image acquired during a CZ crystal growth process. Source: Reproduced by permission of GlobalWafers Co., Ltd.

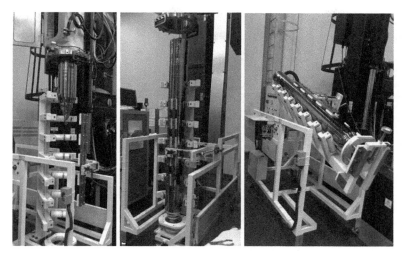

Figure 2.16 Left: lowering a silicon boule from the growth chamber. Center: boule/ingot lowered to the prescribed position and held by the jig. Right: rotating the boule/ingot to be ready for the next steps of wafer forming. Source: Reproduced by permission of GlobalWafers Co., Ltd.

After the crystal growth process is completed, the silicon crystal boule/ingot must be retrieved from the growth chamber for subsequent processing. The boule/ingot retrieving procedures are shown in Figure 2.16 with the equipment and jig. First, the grown silicon ingot is moved and lowered from the growth chamber, as shown in Figure 2.16(left). Next, the ingot is lowered to a position relative to a jig designed to hold the ingot, as in Figure 2.16(center). After that, the jig will pick up the ingot to rotate and/or configure it for transporting to the next steps of wafer forming.

2.7.2 Wafer Forming

The long silicon boule/ingot must be cropped into manageable lengths for wafer manufacturing. Figure 2.17 shows an equipment for slicing long boule/ingot into segments of ingots for wafer manufacturing. After that, the cylindrical surfaces of ingots must be trimmed into the prescribed size (for example, diameters of 200 mm or 300 mm). The ingot surface grinding process is employed to trim the cylindrical face of ingots. Figure 2.18 illustrates an ingot before and after trimming by using surface grinding equipment.

After surface trimming, the crystalline orientation of ingots must be characterized and marked for both wafer slicing and microelectronic fabrication processes. Crystalline orientation can be characterized by various methods, including (i) optical etching, (ii) X-ray/electron diffraction, (iii) SEM or TEM [Kikuchi (1990); Schwarzer (1997)]. The X-ray diffraction method, widely used because of its high precision and ease in handling, can be further divided into the diffractometer method, the Laue method, and others [Kikuchi (1990)]. X-ray diffraction equipment is illustrated

Figure 2.17 Boule/ingot is sliced into segments of manageable length for wafer manufacturing. Source: Reproduced by permission of GlobalWafers Co., Ltd.

Figure 2.18 Left: silicon ingot before trimming using surface grinding. Right: trimmed ingot after surface grinding. Source: Reproduced by permission of GlobalWafers Co., Ltd.

in Figure 2.19. When the crystalline orientation is properly measured and characterized, the ingot will be furnished with an orientation flat (200 mm diameter or smaller) or an orientation notch (300 mm) on the cylindrical face of the ingot for the identification of crystalline orientation, necessary in the microelectronics fabrication processes.

The next step in wafer forming is the slicing of crystal ingot into wafers through "wafer slicing" processes. Wafer slicing can be done using equipment such as ID saws or wiresaws, which will be discussed in more details in Chapters 4–6. Figure 2.20(left) shows an industrial slurry wiresaw for wafer slicing. During the slicing process, the ingot is typically mounted on a ceramic base that allows the wire web of the slurry wiresaw to slice through the entire cross-section of the ingot while holding the sliced wafers in place by epoxy glue which affixes the sliced wafers to this ceramic base. After wafer slicing is completed, the epoxy glue is softened by elevating the temperate to remove each wafer. These as-sliced wafers will then go through a post-slicing cleaning process. Figure 2.20(center) shows an equipment set-up for post-slicing cleaning of as-sliced wafers. After that, the as-sliced wafers go through a process of edge rounding, or edge profiling, as shown in Figure 2.20(right), before the lapping and polishing processes. Silicon is a brittle material and is prone to crack formation and crack propagation if there are chips or tiny cracks on the edge of the wafer. Edge rounding of wafers reduces the likelihood of stress concentration

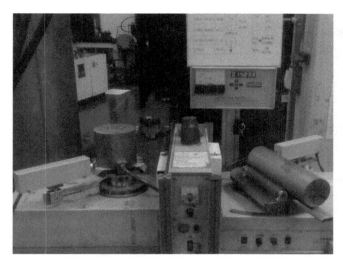

Figure 2.19   X-ray diffraction equipment for the measurement of crystalline orientation, and marking by an orientation flat or notch. Source: Reproduced by permission of GlobalWafers Co., Ltd.

Figure 2.20   Wafer forming. Left: An industrial slurry wiresaw for slicing an ingot into wafers. Center: equipment for the cleaning of as-sliced wafers after wafers are removed from the ingot mount. Right: equipment for edge rounding of individual wafers. Source: Reproduced by permission of GlobalWafers Co., Ltd.

and crack propagation that can damage wafers during handling and subsequent processing of wafer manufacturing, as described in Section 2.4.8 with more detail.

2.7.3   Wafer Lapping and Polishing

Referring to Figure 2.1, the "wafer polishing" operation after "wafer forming" includes a plethora of processes in lapping, polishing, cleaning, and inspection. These processes in an industrial establishment will be discussed in the following with illustrations.

Lapping

Lapping removes a layer of wafer surface with subsurface cracks and damage induced in the mechanical wafer slicing process to produce an initial global planarization on wafer surfaces. Figure 2.21(left) shows a lapping machine with five lapping plates, each holding a multiple number of wafers. These wafers are placed

Figure 2.21 Wafer polishing. Left: a lapping machine that processes multiple wafers on five lapping plates. Right: equipment for post-lapping cleaning. Source: Reproduced by permission of GlobalWafers Co., Ltd.

in between two counter-rotating plates, fed with a slurry consisting of abrasive grits (typically alumina with small grit size). Lapping improves the wafer surface characteristics of smoothness and the flatness. The total thickness variation (TTV) after the lapping process is typically close to 1 μm.

However, lapping in itself is a mechanical free abrasive machining (FAM) process, although with much less brutal force than the slicing process, which also creates subsurface damage even though at a much smaller scale. The acid etching process removes this small layer of damage [Kulkarni and Erk (2000)]. Figure 2.21(right) shows the equipment for cleaning after lapping. This layer of post-lapping damage is removed by basic chemical or acid etching, producing a surface that is nearly free of cracks, damage, and crystallographical flaws. Ultrasonic cleaning and a very brief etching are often utilized to remove all particles, organic, and inorganic films from the wafer surface before etching.

Post-lapping and Pre-polishing

Various equipment and processes are employed after lapping and before final polishing. Figure 2.22 illustrates these post-lapping and pre-polishing equipment and processes. Note that such processes may vary from manufacturer to manufacturer.

Acid etching is performed for wafers after the lapping and cleaning process using an equipment shown in Figure 2.22(a). Post-etching inspection for the surface characteristics of wafers, using the ADE 7000 equipment shown in Figure 2.22(b).

Equipment is shown in Figure 2.22(c) to perform the blast process on the back side of wafers. Figure 2.22(d) shows back side cleaning. Figure 2.22(e) shows the equipment to administer edge polishing to removes cracks and rough edges in order to prevent crack formation and propagation due to mechanical stress concentration incurred in the subsequent processes and in microelectronic fabrication. Figures 2.22(f) and (g) show the equipment used for pre-poly back cleaning and polyback processes, respectively. The equipment in Figure 2.22(h) is for pre-low-temperature oxide (LTO) cleaning, with LTO processing following it.

There are two branches of options for the process of oxide strip: (i) edge oxide strip, and (ii) protective tape for the edge oxide strip. Figure 2.22(i) shows an

Figure 2.22 Post-lapping to pre-polishing equipment and processes: (a) acid etching, (b) post-etching inspection, (c) blast of back surface, (d) back side cleaning, (e) equipment for edge polishing, (f) pre-poly back cleaning, (g) polyback, (h) pre-low-temperature oxide (pre-LTO) cleaning equipment and LTO, (i) protective tape edge oxide strip, (j) inspection before the polishing process, (k) pre-mounting cleaning, and (l) mounting for polishing. Source: Reproduced by permission of GlobalWafers Co., Ltd.

equipment for the protective tape for edge oxide strip. Figure 2.22(j) shows the equipment for inspection before polishing process. The equipment in Figure 2.22(k) performs pre-mounting cleaning before the polishing process. Figure 2.22(l) shows the final mounting for polishing.

The cleaning of the backside surface of the wafer is important in the lapping/polishing process [Chou et al. (2005); Park and Sohn (2012); Ruzyllo and Novak (1994)]. This is demonstrated in the processes at various stages of cleaning in Figure 2.22, especially before the final polishing processes to produce prime wafers. Impurities on silicon wafers can be categorized as (i) particles and films, including molecular compounds, ionic materials, and atomic species, and (ii) absorbed gases that are of little practical consequence in wafer processing [Kern (1990)]. Figure 2.23 demonstrates that even small particles on the backside of a wafer can cause deformation of the wafer surface exposed to the light of the lithography process that can result in focus spot failure or distortion due to depth of focus or reduced process window, especially when the feature size is decreased [Balu et al. (2018)].

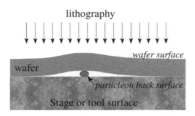

lithography

Figure 2.23 Illustration of the negative impacts on lithography with particles on the backside of a wafer.

Figure 2.24 Wafer polishing equipment: Polishing of wafers, with a zoom-in view on the upper-right corner for illustration of details. Source: Reproduced by permission of GlobalWafers Co., Ltd.

Polishing

The last step of the wafer polishing process is the polishing of wafer surface to produce prime wafers, ready for microelectronics fabrication. This is illustrated in Figure 2.24 in which an industrial polishing equipment performs the final polishing of wafers to produce prime wafers. The zoom-in view at the upper-right corner of the photo provides a more detailed picture of the apparatus for wafer polishing.

2.7.4 Wafer Preparation

Finally, the polished prime wafers are cleaned, inspected and packaged in a "boat" ready for shipping.

Figure 2.25(left) shows equipment for pre-cleaning after wafer polishing. The equipment in Figure 2.25(center) performs the flatness sorting of prime wafers. The system in Figure 2.25(right) performs the final cleaning – the RCA cleaning (see also Sections 9.5 and 9.6).

Figure 2.26(left) shows the inspection of the appearance of wafers. The equipment in Figure 2.26(center) performs particle scanning. Figure 2.26(right) shows packaging of prime wafers in a "boat" that is ready for shipping and

Figure 2.25 Wafer preparation. Left: pre-cleaning after the polishing process. Center: flatness sorting. Right: final cleaning. Source: Reproduced by permission of GlobalWafers Co., Ltd.

Figure 2.26 Wafer preparation. Left: appearance inspection. Center: particle scanning. Right: packaging in wafer boats, ready for shipping. Source: Reproduced by permission of GlobalWafers Co., Ltd.

micro-electronic fabrication. This is the very last step of wafer manufacturing after which the prime wafers are ready to be used in micro-electronic fabrication.

2.8 Summary

In this chapter, the process flow of wafer manufacturing is introduced – from crystal growth to prime wafers. Individual processes in the four categories of wafer manufacturing or wafer production, namely, (i) crystal growth, (ii) wafer forming, (iii) wafer polishing, and (iv) wafer preparing, are discussed. The melt growth and vapor growth are most commonly used in growing single and polycrystalline semiconductor crystals, and are discussed with relevant details, including the CZ and FZ growth methods, as well as epitaxial growth. Processes of wafer forming from ingots are presented, including cropping, trimming, orientation identification, slicing with different machine tools, and edge rounding. After that, the lapping, grinding, and polishing processes associated with wafer polishing are introduced. The industrial practice of wafer manufacturing is presented with illustrations of wafer manufacturing equipment and processes that can vary from manufacturer to manufacturer. Finally, prime wafers are prepared with RCA cleaning, inspection, and packaging to complete the wafer manufacturing processes. These processes of wafer manufacturing are further discussed in the following chapters of this book.

References

Balu E, Tseng WT, Jayez D, Mody J and Donegan K 2018 Wafer backside cleaning for defect reduction and litho hot spots mitigation *29th Annual SEMI Advanced Semiconductor Manufacturing Conference (ASMC).*

Bhagavat M, Prasad V and Kao I 2000 Elasto-hydrodynamic interaction in the free abrasive wafer slicing using a wiresaw: Modeling and finite element analysis. *Journal of Tribology* **122**(2), 394–404.

Canfield PC and Fisk Z 1992 Growth of single crystals from metallic fluxes. *Philosophical Magazine B* **65**(6), 1117–1123.

Chou WY, Tsui BY, Kuo CW and Kang TK 2005 Optimization of back slide cleaning process to eliminate copper contamination. *J. Electrochem. Soc.* **152**(2), G131–G137.

Cröll A and Volz MP 2009 Detached Bridgman growth–a standard crystal growth method with a new twist. *MRS Bulletin* **34**(4), 245–250.

Dhanaraj G, Byrappa K, Prasad V and Dudley M 2010 *Springer Handbook of Crystal Growth*. Springer.

Feigelson RS 2004 *50 Years Progress in Crystal Growth–A reprint collection*. Elsevier, Amsterdam, the Netherlands.

Fickett B and Mihalik G 2001 Multiple batch recharging for industrial CZ silicon growth. *Journal of Crystal Growth* **225**, 580–585.

Fisk Z and Remeika JP 1989 *Handbook of the Physics and Chemistry of Rare Earths* Elsevier Amsterdam chapter 12, p. 53.

Gise P and Blanchard R 1986 *Modern Semiconductor Fabrication Technology*. Prentice Hall.

Glassbrenner CJ and Slack GA 1964 Thermal conductivity of silicon and germanium from 3k to the melting point. *Physical Review* **134**(4A), A1058–A1069.

Hurle DTJ 1994 *Handbook of Crystal Growth - 2: Bulk Crystal Growth - Part A: Basic Techniques*. North-Holland.

Jurisch M and Eichler S 2003 The development of LEC technology for GaAs single crystal growth from laboratory sacel to mass production *Czochralski Symposium 2003*, Torun/Kcynia, Poland.

Kao I 2004 Technology and research of slurry wiresaw manufacturing systems in wafer slicing with free abrasive machining. *the International Journal of Advanced Manufacturing Systems* **7**(2), 7–20.

Kern W 1990 The evolution of silicon wafer cleaning technology. *J. Electrochem. Soc.* **137**(6), 1887–1892.

Kikuchi T 1990 Single crystal orientation measurement by x-ray methods. *The Rigaku Journal* **7**(1), 27–35.

Krause M, Friedrich J and Muller G 2003 Systematic study of the influence of the Czochralski hot zone design on the point defect distribution with respect to a perfect crystal. *Materials Science in Semiconductor Processing* **5**, 361–367.

Kulkarni MS and Erk HF 2000 Acid-based etching of silicon wafers: Mass-transfer and kinetic effects. *Journal of the Electrochemical Society* **1**(147), 176–188.

Li J, Kao I and Prasad V 1998 Modeling stresses of contacts in wiresaw slicing of polycrystalline and crystalline ingots: Application to silicon wafer production. *Journal of Electronic Packaging* **120**(2), 123–128.

Lu Z and Kimbel S 2011 Growth of 450 mm diameter semiconductor grade silicon crystals. *Journal of Crystal Growth* **318**(1), 193–195.

Mullin JB 1993 *Handbook of Crystal Growth–Bulk Crystal Growth: Basic Techniques* vol. II Elsevier chapter 3, pp. 105–130.

Mullin JB, Straughan BW and Brickell WSJ 1965 Liquid encapsulation techniques: The use of an inert liquid in supressing dissociation during the melt-growth of InAs and GaAs crystals. *Phys. Chem. Solids* **26**, 782–784.

Noghabi OA, MHamdi M and Jomaa M 2011 Effect of crystal and crucible rotations on the interface shape of Czochralski grown silicon single crystals. *Journal of Crystal Growth* **318**, 173–177.

Pamplin BR 1980 *Crystal Growth* 2nd edn. Pergamon Press, Oxford.

Park CG and Sohn HS 2012 Simultaneous removal of particles from front and back sides by a single wafer backside megasonic system, vol. **187**, pp. 167–170.

PCA 2014 Raw material process, by the Polishing Corporation of America. URL http://www.pcasilicon.com/silicon-wafers/raw-material-process.

Plummer JD, Deal MD and Griffin PB 2000 *Silicon VLSI Technology Fundamentals, Practice and Modeling*. Prentice Hall, Upper Saddle River, NJ.

Ruzyllo J and Novak RE 1994 Cleaning technology in semiconductor device manufacturing In *Proceedings of the Third International Symposium* (ed. Ruzyllo J and Novak RE), vol. 94–7.

Schwarzer RA 1997 Automated crystal lattice orientation mapping using a computer-controlled sem. *Micron* **28**(3), 249–265.

SVM 2014 Silicon wafer manufacturing semiconductor process, by the Silicon Valley Microelectronics, Inc. URL http://www.svmi.com/silicon-wafer-manufacturing-semiconductor-process.

Wagner RS and Ellis WC 1964 Vapor-liquid-solid mechanism of single crystal growth. *Applied Physics Letters* **4**(5), 89–90.

Wang CL, Zhang H, Wang TH and Ciszek TF 2003 A continuous Czockralski silicon crystal growth system. *Journal of Crystal Growth* **250**, 209–214.

Wang CL, Zhang H, Wang TH and Zheng LL 2006 Solidification interface shape control in a continuous Czochralski silicon growth system. *Journal of Crystal Growth* **287**(2), 252–257.

Wells AF 1946 Crystal growth. *Annual Reports to the Progress of Chemistry* pp. 62–87. Chemical Society, London.

Wood A, Schrock E and Grigg F 2013 Microelectronic device wafers and methods of manufacturing. US Patent 8,455,983.

3

Process Modeling and Manufacturing Processes

3.1   Introduction

This chapter focuses on the process modeling of wafer manufacturing. Wafer manufacturing processes as presented in Chapter 2 involve brittle crystal materials. This chapter will be built upon the subjects of forming in Section 2.4 and polishing in Section 2.5, to introduce the modeling of manufacturing processes of brittle semiconductor materials. The characteristics of brittle materials, such as semiconductors, are different from those of ductile materials, such as metals, which further differentiates process modeling. Different wafer manufacturing processes also incur specific levels of interaction between the tools and materials. In addition, most wafer manufacturing processes are abrasive-based, which can be broken into either (i) a bonded abrasive manufacturing process (BAM), or (ii) a free abrasive manufacturing process (FAM). Each process will be discussed in more detail, with comparison, in this chapter. The rolling-indenting and scratching-indenting processes for brittle materials will be introduced along with other related issues in wafer manufacturing processes, such as recovering and recycling abrasives in the wiresawing process.

3.2   Wafer Manufacturing and Brittle Materials

Semiconductor materials are brittle materials, which include silicon, silicon carbide, III–V compounds, II–VI compounds, and other optoelectronic materials. The wafer forming and polishing processes of semiconductors employ various manufacturing processes for the removal of materials in order to produce prime wafers for fabrication, as discussed in Chapter 2. The processes include slicing, grinding, lapping, and polishing. In this section, the mechanical aspects of modeling and machining models, namely ductile and brittle machining, are discussed, with a presentation of future technology and research on more efficient wafering technology.

Brittle materials are characterized by having little deformation or strain, usually less than 5% of strain in a tensile test, when subject to stress [Callister and Rethwisch (2014); Kakani and Kakani (2004)]. Brittle materials often fail in an abrupt manner, absorbing relatively little energy before fracture, as indicated by the area under the

Wafer Manufacturing: Shaping of Single Crystal Silicon Wafers,
First Edition. Imin Kao and Chunhui Chung.
© 2021 John Wiley & Sons Ltd. Published 2021 by John Wiley & Sons Ltd.

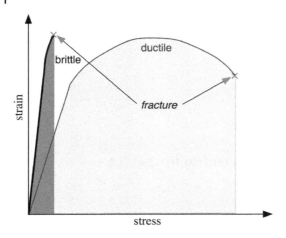

Figure 3.1 Typical stress–strain curves for brittle and ductile materials. Brittle materials show little strain (< 5%) before fracture, and also absorb much less energy before the fracture, as indicated by the areas under the curve.

curve in Figure 3.1. A comparison of the stress–strain curves between brittle materials, considered here for most crystal materials in wafer manufacturing, and typical ductile materials, such as most metals, is illustrated in Figure 3.1.

Conventional wafer processing begins with the slicing of a crystalline ingot using inner-diameter (ID) saws or wiresaws [Bhagavat et al. (2010); Chonan and Hayase (1987); Chonan et al. (1993); Chung and Kao (2008); Clark et al. (2003a,b); Kao (2004); Kao and Bhagavat (2007); Kao et al. (1998a); Kao et al. (2010); Kojima et al. (1990); le Scanff (1988); Li et al. (1998); Mech (1974); Sahoo et al. (1998); Shimizu (1976); Wells (1987, 1985)], followed by lapping and/or grinding and polishing. The slicing, lapping, and polishing of brittle wafer materials are "brittle machining processes," utilizing either bonded abrasive grits on machine tools or free abrasives in rolling and indenting. The common trait of such brittle machining processes is the subsurface radial and transverse cracks formed during the machining processes, as shown in Figure 3.2(b).

After slicing (Section 2.4), a layer of crystal materials with subsurface damage (SSD) and micro-level surface topography (cf. Figure 1.9) will be further processed using lapping and/or grinding processes (Section 2.5). Finally, etching and chemical-mechanical polishing (CMP) processes are utilized to make prime wafers with mirror-surface finish. The post-slicing and wafer forming processes involve a series of wafer polishing processes, with objectives of

(i) The removal of surface waviness and warp due to slicing which will inevitably cause further subsurface damage (SSD).
(ii) The removal of all SSD to avoid defects when fabricating micro-electronic and other devices on wafer surface.
(iii) The preparation of wafers with a defect-free and mirror-finish surface ready for device fabrication.

The total thickness variation (TTV) of sliced wafers can range from 10–30 microns with SSD ranging from a few microns to tens of microns, depending on the process and control [Kao (2004); Li et al. (1998)]. This results in removing a depth of

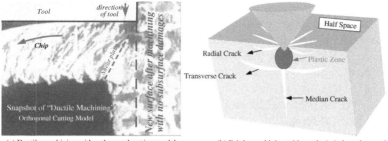

(a) Ductile machining with orthogonal cutting model   (b) Brittle machining with cracks in indentation model

Figure 3.2 (a) Ductile machining: high-speed photo of ductile machining with orthogonal cutting model, without subsurface cracks and damage (Source: Adapted from the SME videos). (b) Brittle machining, with abrasive machining through indentation, such as wafer slicing, lapping, grinding, and polishing with SSD. (Sources: Bhagavat et al. (2000); Lawn (1993); Lawn and Evans (1980); Lawn and Swain (1975); Zhu and Kao (2005).)

20–70 microns of material (about 6–10% of typical wafer thickness) after the wafer is sliced, incurring a loss of material and productivity due to the time required for surface machining – a factor that is manifested by the growth of wafer size, currently at 300 mm and to become 450 mm in the near future [Capraro (2013); Intel News Release (2008); Watanabe and Kramer (2006)]. Thus, it is very important to understand the mechanisms of machining and modeling in order to improve the process so as to be more effective and economical in materials processing and manufacturing.

3.3 Ductile Machining Versus Brittle Machining

Two fundamentally different machining models are illustrated in Figure 3.2, namely, the ductile machining model and the brittle machining model. When ductile materials are removed from the surface of bulk in a conventional machine tool, such as a lathe or a milling machine, the machining is dominated by shear in a process commonly known as the *orthogonal machining model*, as illustrated in Figure 3.2(a) [Groover (2012); Kalpakjian and Schmid (2013)]. The cutting tool moves with a large force to cause the ductile material to be separated from the workpiece bulk, due to shear stress at the contact interface between the tool and the material, generating a shear plane that marks the transition of the unaffected bulk surface from the chip. A corrugated chip is removed from the bulk surface, and flies off the surface of the tool. A snapshot of the orthogonal machining in action during a ductile machining process is illustrated in Figure 3.2(a). The ductile machining, based on the orthogonal model, results in an exterior surface free of subsurface cracks and damage.

Brittle machining, on the other hand, can be performed with material removal by shear on the surface with a tool, accompanied by a large number of cracks and fractures caused by the shearing process, both on the exposed surface and beneath the surface. Brittle machining can also be accomplished by surface indentation using indenters with concatenation of subsurface cracks that disengage the free chips on the surface layers from the bulk.[1] When abrasive grits press onto the surface of brittle materials in a "loading process," they act as indenters. As shown in Figure 3.2(b), when an indenter with higher hardness indents onto a brittle material, a plastic zone is formed immediately under at the tip of indenter. When the indenter applies loading, the median cracks start to form, as indicated in Figure 3.2(b), which can penetrate deep into the workpiece bulk along the loading direction. When the indenter unloads in retreating movement, the transverse cracks and radial cracks start to form, which runs nearly parallel to the surface of the bulk material at a more shallow level beneath the surface [Lawn (1993); Lawn and Evans (1980); Lawn and Swain (1975)]. Such loading and unloading create a very unique crack formation and propagation in brittle materials. Such phenomenon of crack formation during the phases of loading and unloading can take place in both single and polycrystalline semiconductor materials considered in this book.

When comparing the two machining models, the brittle machining model does not generate continuous chips that characterize the ductile machining using the orthogonal cutting model. The distinction in machining process modeling between ductile and brittle machining is important in wafer manufacturing for slicing, forming, lapping, polishing, removing the surface waviness and warp, and the formation of subsurface damage (SSD).

3.4   Abrasive Machining in Wafer Manufacturing

In this section, the models of abrasive machining will be presented in the context of various processes of wafer manufacturing. Most wafer manufacturing processes utilize abrasives to remove materials. These abrasives, whether loose/free or bonded, are the tools of machining.

Abrasive machining encompasses a collection of various machining processes in which material is removed from a workpiece using abrasive grits. Figure 3.3(a) illustrates the schematics of a BAM process. Machining ductile materials with abrasives is often employed to render tighter tolerance. Examples of this type of machining include grinding, honing, and polishing. A photo of a precision grinding wheel machine tool is shown in Figure 3.3(b). Abrasive machining processes for such purposes are usually more expensive, with very small amount of material removal. Brittle materials, especially the semiconductor materials, are often machined by employing a plethora of abrasive machining processes.

---

1 The former is often accomplished by bonded abrasives tools that apply more brutal force to remove brittle materials from the surface. The latter is often accomplished through loose or free abrasives in a three-body abrasion process. This abrasive-based machining of brittle materials will be discussed further in Section 3.4.

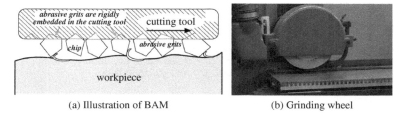

(a) Illustration of BAM   (b) Grinding wheel

Figure 3.3 (a) An illustration of the bonded abrasive machining (BAM) process; the schematic shows the abrasive grits bonded rigidly into the tool, such as a grinding wheel or the diamond-coated blade of an ID saw, which carries the abrasive grits for surface scratching and machining. (b) Photo of a grinding wheel machine tool for precision machining. Source: Imin Kao.

Abrasive machining processes can be broken into two categories based on the arrangement of abrasive grits as the agent of machining [Chung and Kao (2008a)]:

1. Bonded abrasive machining (BAM)
2. Free abrasive machining (FAM).

More detail will follow in the next sections.

3.4.1 Bonded Abrasive Machining (BAM)

In a bonded abrasive machining process, the abrasive grits are bonded or attached to the machine tool to impose abrasive action on the workpiece and remove materials from the workpiece. Among the wafer manufacturing processes, ID sawing is a BAM process. The ID saw is stretched at a very high tension with abrasive grits (typically diamond grits) attached to the edge of the hollow center of the ID saw's metal plate. The abrasives remove material from the ingot, similar to a ploughing process. In addition, a reciprocating saw operated with a diamond-impregnated wire is also a BAM process. The impregnated diamond grains on the circumferential surface of the wire act as the agent of machining in a ploughing process, typical of the BAM process. Another example is the wafer surface grinding process.

Such BAM processes usually have higher material removal rate, but involve brutal force, primarily in shear, to remove material. What accompanies such processes is severe crack formation, in both lateral and radial directions, beneath the surface left by the BAM process, in addition to surface fractures and scratch marks. Although the surface fractures and indentations can help remove the material, it is not easy to control the process to reduce surface roughness and subsurface damage, which are not permitted in wafers produced for the semiconductor industry.

3.4.2 Free Abrasive Machining (FAM)

In a free abrasive machining process, the abrasive grits are not attached to the machine tool, but rather free to roll or move in "three-body abrasion," as illustrated in Figures 3.4 and 5.3. The "three body" illustrated in the schematic drawing

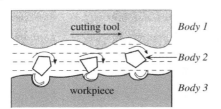

Figure 3.4 Schematic of the "three-body-abrasion" process showing the cutting tool as body 1, the free abrasive grits as body 2, and the workpiece as body 3, moving with respect to one another without a physical bonding among them.

in Figures 3.4 refers to the tool, workpiece surface, and the abrasive grits as three distinct entities, with motions relative to one another, and without physical bonding among them. The cutting tool is typically the primary moving part of the process that carries the abrasive grits in slurry fluid. The workpiece is typically held in a stationary configuration, although it can also move. The abrasion stems from the relative motion among the three bodies, while the unconstrained free abrasive grits roll, indent, and scratch the surface of the workpiece bulk, as well as the tool. The abrasion taking place between the free abrasive grits and the surface of the workpiece causes materials to be removed from the substrate surface through the rolling, indenting, and scratching actions of the abrasive grits at the contact interface, as illustrated in Figures 3.4.

The manufacturing process modeling and mechanisms of FAM will be discussed in more detail in Section 3.7.

Because of the demands of prime wafers for semiconductor microelectronics fabrication, BAM and FAM processes such as ID sawing, wiresawing, lapping, surface grinding, and polishing have been widely employed for the purposes of forming wafers. Among the wafer manufacturing processes, wiresawing, lapping and polishing are examples of FAM processes. In the case of wiresawing, the fast-moving wire is the tool that interacts with abrasive slurry to create hydrodynamic pressure to force abrasives to make contact with the surface of brittle materials for machining. In the case of lapping and polishing, the abrasive grits move freely and independently. They must be forced onto the workpiece, using another object like a polishing cloth or a lapping plate, to remove the material from the surface of the workpiece. In addition, the slurry of the slow-moving lapping and polishing processes also works as a lubricant and coolant, and takes away fine kerf materials that come off the surface of the workpiece. However, the abrasive slurry is essential for machining in the wiresawing FAM process that involves hydrodynamics, using a mechanism which is different from the FAM processes in lapping and polishing. This will be elaborated in Section 5.5.

3.5 Abrasive Materials

Abrasive materials are essential in both BAM and FAM processes and can be classified as conventional abrasives or superabrasives. Conventional abrasive materials, such as alumina oxide and silicon carbide, have been used by humans for shaping tools and machining. They are found naturally on the earth, and are often abundant and cheap. Superabrasives, such as cubic boron nitrides (CBN) and industrial diamonds, are mostly synthetic and are made much harder than readily available conventional abrasives.

The abrasive materials are typically fine particles or grains, which are also called abrasive grits. The grain or grit size is measured using a screen mesh procedure. Grit size is defined by the distribution of grit retained on a sieve setup that meets the requirements. The grit sizes are related to the number of lines per inch (25.4 mm) length of each sieve, for example, 150 lines per inch. Employing this sieve with 150 lines per inch, abrasive grits 150 will pass. Thus, abrasive grits with smaller grain sizes will have a larger number. There are several standards for classifying the grain size of abrasive materials. The standards of the classification of abrasive sizes and hardness will be discussed in the next sections.

3.5.1 Classification of the Grain Size of Abrasive Materials

There are various international standards to classify the grain (or grit) size of abrasive materials, including the standards defined by

- ANSI: the American National Standards Institute.
- FEPA: the Federation of European Producers of Abrasives, or the Fédération Européenne des Fabricants de Produits Abrasifs.
- JIS: Japanese Industrial Standard, issued by the Japanese Standards Association.

The three international standards have different procedures and definitions, and are broadly grouped into (i) macrogrit and (ii) microgrit, based on the range of sizes. The three systems are not strictly comparable. Adding to the complication is the different numbering system among the three main standards. For example, the number 400 grain size may refer to different sizes of grit in ANSI, FEPA, and JIS standards[2]. In general categorization, the abrasives can be regarded as "bonded abrasives" that are typically relevant to grinding wheels and sandpapers, and "coated abrasives" that are used in lapping, polishing, and cleaning. In the context of wafer manufacturing, the "microgrits," which have a grit size of less than 50 microns[3], are of primary interest. Useful information can be found in the following sources for further reading FEPA (2014); Fine Tools (2014); UAMA (2014); Washington Mills (2014). A useful comparison of grit sizes can be downloaded as a PDF file compiled by Smokintbird [Smokintbird (2009)].

The American National Standard defines several standards related to abrasive materials. Specifically, the ANSI-B74.18-2006 and ANSI-B74.12-2009 document defines the standard coated abrasive materials and the specifications for the size of abrasive grains [ANSI (2006, 2009)]. This project was a collaboration with the ASTM and then the Coated Abrasive Manufacturers Institute (CAMI)[4], and has

---

2 As an example, the 400 numbering of the three standards represent the following average particle sizes: the ANSI 400 is 22 microns; the FEPA P400 is 35 microns; the FEPA F400 is 17 microns; the JIS 400 (new) is 30 microns; and JIS 400 (old) is 34 microns.
3 One micron, μm, is the same as one micrometer, which is 1×10^{-6} meter in SI units.
4 The Unified Abrasives Manufacturers' Association was formed in 1999 from the merging of four predecessor organizations: the Abrasive Grain Association, Coated Abrasives Manufacturers' Institute, Diamond Wheel Manufacturers' Institute and the Grinding Wheel Institute.

gone through several revisions, with the last revision being the ANSI B74.18-1996. This ANSI standard considers the particle size distribution as "the range of the particle sizes within a grit size as expressed by the cumulative distribution graph of micron size vs. % passing based on volume for microgrits or by the percentages of overgrade, control and fines determined by screen grading" [ANSI (2006)]. An "electrical resistance method" for testing microgrit is classified in [ANSI (2006)]. For example, one can determine number and size of particles suspended in an electrically conductive liquid by flowing the suspension through a small aperture between electrodes. The standard specifies the size distribution at d3 (3%), d50 (50%), and d94 (94%) values in a typical Gaussian distribution.

The size distributions are defined by FEPA, which distinguishes between grains for sanding paper (FEPA P) and grain for sharpening stones or wheels (FEPA F). FEPA defines a grade by defining a range of grain sizes. For example, by the FEPA standard for microgrit grade F220, no more than 3% of grains can be larger than 75 microns, at least 50% must be in the range 50.0 to 56.0, and at least 94% must be larger than 45 microns. The most commonly used standard for abrasives in wafer manufacturing is FEPA F. FEPA P is often used for sanding paper.

JIS, the Japanese Industrial Standard, has an old and a new standard. The JIS standards were formed by the Industrial Standardization Law enacted in 1949. The Industrial Standardization Law was revised in 2004 and began implementation on 1 October 2005. The new JIS standards were adopted within a three-year transition period, after 1 October 2008.

Relevant abrasive grades of microgrit with particle sizes, based on the three standards, are listed in Table 3.1. Note that the new ANSI standard specifies a range of values for each grit designation at d3 (3%), d50 (50%), and d94 (94%) values. The ANSI grit sizes listed in Table 3.1 are based on the average values of the d50 range from the new ANSI B74.10.2010 [ANSI (2010)], different from the previous standard of B74.10-2001. Table 3.2 lists the standard micron values for microgrits measured by Electrical Resistance Apparatus, as defined in the new ANSI B74.10.2010.

3.5.2 Hardness of Abrasive Materials

The most important property of an abrasive grits is the hardness. Abrasives having higher hardness are suitable for BAM or FAM processes. Abrasives must be significantly harder than the workpiece material in order for them to be effective in machining. The hardness of materials can be measured in many different standards. The "Knoop hardness" is often employed for quantitative comparison of hardness among typical abrasive materials. The "Mohs hardness" is a rough scale of hardness measure based on the resistance of a smooth surface to scratching, devised in the 19th century to determine the hardness of a mineral.[5] Table 3.3 lists abrasive materials and a few common materials for comparison of their Knoop hardness and

---

5 The Mohs hardness scale has a range from 1 to 10, with talc being 1 and diamond being 10. Aluminum oxide (corundum) is listed at 9 (for example, see [Klein (2008)]). Obviously, a lot of synthetic materials have a Mohs hardness between 9 and 10. A more precise and scientific measure of hardness is the Knoop hardness for ceramic materials, as listed in Table 3.3.

Table 3.1 Some selected average particle sizes of abrasive materials, as defined in ANSI, FEPA, and JIS standards.

ANSI	FEPA P	FEPA F	JIS	Average grit size in μm
–	P280	F230	–	53
240	–	–	–	50
–	P360	–	320	40
280	–	–	–	39
320	P500	–	400	30
–	–	F320	–	29
–	–	–	500	25
360	P800	F360	–	22
–	–	–	600	20
400	–	–	–	18
–	–	F400	700	17
–	P1200	–	–	15
500	–	–	800	14
–	–	–	1200	9.5
–	–	F600	–	9.3
–	–	–	1500	8
800	–	–	–	7.5
–	–	F800	–	6.5
–	–	F1000	–	5
1200	–	–	–	3.8
–	–	F1200	4000	3
–	–	F1500	–	2

Note: The ANSI sizes are based on the d50 average of the new B74.10.2010, different from the previous standard.

Mohs hardness. Materials with a range of hardness numbers, shown in Table 3.3, indicate that hardness can vary depending on process of manufacturing; other materials are shown with a typical hardness number. The following abrasive materials are commonly used in wafer manufacturing: aluminum oxide, silicon carbide, boron carbide, and diamond.

Table 3.2 ANSI standard size for microgrits, based on B74.10.2010 [ANSI (2010)].

Grit designation	d3 value	d50 value	d94 value
240	74–80	47–53	26–32
280	67–72	37–42	20–25
320	55–60	27–32	15–19
360	38–42	21–25	10.5–14.5
400	31.5–35.0	16.5–20.0	7.5–10.5
500	28.0–30.5	12.5–15.3	5.5–8.0
600	24.0–27.0	9.2–11.9	–
800	20.5–22.5	6.5–8.8	–
1000	16–18	5.2–6.4	–
1200	13.5–15.5	2.8–4.8	–

Table 3.3 Knoop hardness and Mohs hardness of selected abrasive materials.

Abrasive material	Knoop hardness (kg mm^{-2})	Mohs hardness
common glass	350–500	5.5–6.5
Silicon	1100	7
Flint, quartz	800–1100	**7**
Tungsten carbide	1800–2400	9–9.5
Titanium nitride	2000	9
Titanium carbide	1800–3200	9–9.6
Aluminum oxide	2100	**9**
Silicon carbide	2500	9.3
Boron carbide	3200	9.7
Cubic boron nitride	4500	9.9
Diamond	7000	**10**

3.5.3 Commonly Used Abrasive Materials in Wafer Manufacturing

Commonly used abrasive materials in wafer manufacturing include silicon carbide, aluminum oxide, boron carbide, and diamond. These abrasive materials are further discussed, as follows.

Silicon Carbide
Silicon carbide (SiC), also known as carborundum (9.3 on the Mohs hardness scale), is a synthetic abrasive material produced by the Lely crystal growth process [Lely (1955)] or by the chemical vapor deposition (CVD) process. Silicon carbide exists in about 250 crystalline forms [Lely (1955); Wikipedia (2014b)]. The alpha silicon carbide (α-SiC), with hexagonal crystal structure, is the most commonly found in the abrasives industries. Other forms of silicon carbide include the beta silicon carbide (β-SiC) and amorphous silicon carbide (a-SiC). Two common types of SiC used in industry include:

- Black silicon carbide: black SiC has a lower purity from 95% to 98% of SiC in content. It is tougher and is used in a wide variety of applications, such as abrasives in the BAM and FAM processes. In general, black SiC is cheaper than green SiC and, hence, more popular in industrial usage.
- Green silicon carbide: this is one of the most pure SiC, typically containing 99% or greater content of SiC. Being more friable than black SiC, it is generally used in grinding wheels for particular grinding properties or in industrial applications requiring a high purity SiC. It is often used as a dressing material to regenerate and expose friable abrasive grits in machining, such as the diamond coated surface of ID saws.

Silicon carbide is most commonly used in wiresawing, as an abrasive material mixed in fluid carrier, such as glycol or water, to form the abrasive slurry for slicing ingots into wafers. See also Section 3.8 and Chapter 4.

Figure 3.5(a) shows an FEM photo of FEPA F400 silicon carbide abrasive grits with a mean size of 17 μm. Figure 3.5(b) shows a photo of FEPA F600 silicon carbide abrasive grits with a mean size of 9 μm.

Figure 3.5 FEM photos of silicon carbide abrasives: (a) FEPA F400 silicon carbide abrasive grits with a mean size of 17 μm, and (b) FEPA F600 silicon carbide abrasive grits with a mean size of 9 μm. Sources: Bhagavat et al. (2005, 2010).

Aluminum Oxides

Aluminum oxide, $Al_2 O_3$, is a chemical compound of aluminum and oxygen, often in its crystalline polymorphic phase α-$Al_2 O_3$. It is commonly called alumina, and can be found naturally on earth in a crystalline form called corundum (9 on the Mohs scale of mineral hardness). When in the form of fine particles, $Al_2 O_3$ can be used as an abrasive due to its high hardness. Aluminum oxide abrasive materials of different particle sizes are often employed in various lapping and polishing processes for forming wafers.

Boron Carbide

Boron carbide, as shown in Table 3.3, has a higher Knoop hardness than both silicon carbide and aluminum oxide. Boron carbide is a competent replacement for silicon carbide, especially in slicing using the wiresaws. However, it is more expensive than silicon carbide or aluminum oxide; therefore, the economic benefit has not been proven to be favorable for boron carbide as a common abrasive material for wafering.

Diamond

Diamond can be found naturally on earth or produced synthetically. Most natural diamonds are formed at extremely high temperature and pressure at depths of 140 to 190 km (87 to 118 miles) in the Earth's mantle [Wikipedia (2014a)]. Diamonds can also be produced synthetically (i) in a high-pressure and high-temperature process that approximately simulates the conditions in the Earth's mantle, or (ii) through the chemical vapor deposition (CVD) process.

Synthetic or natural diamond, in fine particle sizes, is used to form the bonded abrasive coating along the inner ring edge of the inner-diameter saws found in the ID sawing processes for wafer slicing, described in Section 2.4.5, as well as the bonded abrasive coating on the circumference of the diamond impregnated wire of the diamond wire saws in Chapter 6.

3.6 Ductile Machining of Brittle Materials

The surface machining processes of semiconductor wafers have traditionally been carried out by abrasive machining in the brittle machining regime, accompanied by subsurface cracks and damage that needs to be removed to render prime wafers suitable for microelectronics fabrication. Research in the recent years has demonstrated the capability for machining brittle materials (such as silicon, silicon carbide, silicon nitride, etc.) with ductile machining characteristics, commonly known as the orthogonal cutting model, at small depth of cut (50 nm to 2 μm) [Ajjarapu et al. (2004); Blackley and Scattergood (1991); Blake and Scattergood (1990); Fuchs et al. (1986); Morris et al. (1995); Ngoi and Sreejith (2000); Patten et al. (2004, 2005); Patten and Gao (2001); Puttick et al. (1989); Shibata et al. (1996)]. In another free abrasive machining model for brittle material, Chang et al. (2000) took brittle and ductile-regime machining with two-body (bonded abrasive) and three-body (free abrasive) abrasion into consideration. Ductile machining removes materials by means of shear, resulting in no subsurface damage. This is in contrast to the brittle machining mode in which subsurface median, transverse, and radial cracks are created during the indenting and unloading of abrasive grits typical of abrasive machining processes [Lawn and Evans (1980); Bhagavat et al. (2000)].

The fundamental differences between the two machining process models were discussed in Section 3.4 and illustrated in Figure 3.2. In ductile machining with the orthogonal cutting model, the chips come out of the shear plane with the tool advancing on the work piece. The new surface produced by the ductile machining has no subsurface cracks or damage, only surface tool marks. On the other hand, the brittle machining model, subject to the loading and unloading of abrasive indenters, creates subsurface damage and cracks that require subsequent removal processes. When brittle materials are machined in the ductile regime, the surface can be free from cracks, making it possible to achieve global planarization with minimum loss or waste of material.

3.6.1 Research on Ductile Machining and Challenges

A single-point diamond tool (SPDT) has been used in the study of ductile machining in brittle materials [Blackley and Scattergood (1991); Fuchs et al. (1986); Ngoi and Sreejith (2000); Patten and Gao (2001); Puttick et al. (1989); Shibata et al. (1996)]. The signature of having chips coming off the machined surface is used to judge the presence of ductile machining on brittle materials. In a typical fly-cutting process using a SPDT, the machined surface experiences different zones including (i) brittle machining, with visible surface cracks, (ii) ductile machining, with chips of brittle material in a ductile machining mode with no surface cracks, and (iii) transitional zones from ductile machining to brittle machining or vice versa. Although the research using SPDT has successfully demonstrated the capability to perform ductile machining on brittle materials, it is not practical to use an SPDT to ductile-machine a large wafer, such as a 300 mm silicon wafer, due to unfavorable time constraints. For example, it will take hours to finish one pass of cutting on the surface of a single wafer.

3.6.2 Opportunity and Future Research

Various experimental studies have been performed to conduct feasibility studies by employing different tools, machining parameters, and innovative approaches. This is still very much an active research area. However, it will revolutionize the way that wafer surfaces are machined and polished after being sliced from ingots, when and if such ductile machining on brittle materials becomes successful. Because ductile machining does not incur subsurface cracks and damage, the material loss and waste will be reduced to a minimum, resulting in faster and more economical surface machining. Such surface machining technology can also be expanded to other ceramic materials.

The challenge, however, is to understand the fundamental machining mechanism and process parameters, as well as the mechanical and material properties of tools and their interaction with brittle materials in machining. Furthermore, the durability of the tools, speed of machining, total cost, sustainability, and economic impact should be considered when developing such tools for ductile machining. In addition, the machining parameters such as the depth of cut, speed of tool, rake angle, and feed rate are open issues to be investigated. Ultimately, successful pursuit of such research will advance the understanding of the engineering and science of machining, and the wafer manufacturing processes.

3.7 Process Modeling of Wafer Manufacturing Processes

The wafer manufacturing processes outlined in Chapter 2 involves a plethora of individual processes that produce wafers from crystalline ingots. Most machining processes are either BAM or FAM processes, as described in Section 3.4. The schematic illustration of the processes has been introduced in Section 3.4 and illustrated in Figures 3.3 and 3.6.

Free abrasive machining (FAM) processes have been employed for slicing and finishing brittle material such as glass and stones for more than a century. Wafer manufacturing remains very relevant and important because of the demands of prime wafers for semiconductor microelectronics fabrication and silicon wafers for PV applications. FAM processes such as wiresawing, lapping, and polishing are traditionally important manufacturing processes in wafer production. Since the materials in the semiconductor industry are mostly brittle, the FAM process can provide more gentle machining than the BAM process. Various machining theories and models have been developed to understand those processes. In this section, both the BAM and FAM processes in wafer manufacturing will be discussed in conjunction with the theory of brittle material cracking. The two primary wafer forming FAM processes in wafer production, namely the modern slurry wiresaw slicing process and lapping process, will be presented and compared based on manufacturing process models. Additionally, the two main FAM processes will be presented with consideration given to the characterization of manufacturing mechanisms, abrasive grits, and properties of processes.

The bonded abrasive machining (BAM) process in wafer manufacturing was only relevant in ID saws in the past. With the modern slurry wiresaws replacing ID saws at the turn of the century, BAM became less relevant in wafer manufacturing. However, as the size of wafer continues to increase and the needs of surface flattening, removal of surface tool marks and subsurface damage, and global planarization arise, the BAM process of surface grinding of wafers has been brought to light. Surface grinding is utilized to replace or complement the traditional lapping process, especially in the processing of larger wafers, with a greater degree of automation. The surface grinding process will be discussed in Section 9.2 of Chapter 9. The recently emerged diamond wire saws also utilize BAM process which will be discussed in Chapter 6.

3.7.1 Rolling-indenting and Scratching-indenting Process Models of FAM

In the FAM processes, the free abrasive grits may roll and indent the workpiece to remove the material or be embedded into the tool or workpiece and plough

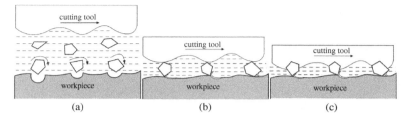

Figure 3.6 Schematic for various processes in FAM. The cutting tool can be a wire or a plate or felt pad depending on the process under consideration. Three different processes are illustrated in (a) rolling-indenting process, (b) rolling-scratching process, and (c) scratching and shearing process. Source: Bhagavat et al. (2005, 2010).

through the workpiece. Schematic drawings of the FAM mechanisms are illustrated in Figure 3.6. According to the size of the abrasive grits and the gap between the tool and workpiece, the material removal mechanisms involved in the free abrasive machining processes essentially can be classified into the three categories as illustrated in Figure 3.6. Since the abrasives are not uniform in size, all of the three free abrasive machining mechanisms will occur to a certain extent during the machining process.

Figure 3.6(a) shows the rolling-indenting process in which the grit size is smaller than the gap between the tool and workpiece. Therefore, the scratching or shearing tends not to happen in this process. Slurry wiresaw machining belongs to this machining category, in which the wire (tool) is separated from making direct contact with the surface of the material through hydrodynamic interaction. Due to the velocity gradient between the material surface and the tool, the slurry moves with different speeds, resulting in the rolling of abrasive grits, indenting onto the material surface with loading and unloading actions. The rolling-indenting mechanism will be further elaborated in Section 5.2 when the modern slurry wiresaw and its mechanism of machining is introduced.

Figure 3.6(b) illustrates the rolling-scratching process in which the grit size is in the same order of magnitude as the gap between the material surface and the tool. Scratching will take place for certain large abrasive grits while the smaller grits roll and indent on the surface. This is most easily seen in the lapping process, in which a lapping plate or pad is pressed against the surface of the wafer with slurry flowing through.

Figure 3.7 illustrates and compares the surfaces of silicon wafers machined under the rolling-indenting and rolling-scratching processes, respectively. Note that the randomly distributed pits on the surfaces of the wafer are the results of abrasive grits indenting on the surface in a random fashion. Figure 3.7(a) shows a photo of the wafer surface, subject to a predominantly rolling-indenting machining, with an abrasive grit embedded on the wafer surface after a FAM process. Figure 3.7(b) shows a photo with the surface morphology after a FAM lapping with a visible scratch that was produced by rolling-scratching action.

(a)   (b)

Figure 3.7 (a) A photo of the surface of a silicon wafer with an abrasive grit embedded on the wafer surface after a FAM process; (b) surface morphology after lapping with a visible scratch. Source: Imin Kao.

Figure 3.6(c) illustrates a FAM process in which the gap is smaller than the average abrasive grits. Under this situation, the abrasive grits can be regarded as being embedded on the tool surface, resulting in a process modeling similar to that in the bonded abrasive machining. However, in this case, the abrasive grits can escape the hold of the tool and become free abrasives again, even though they may be temporarily held. The free abrasive machining under this situation will work like bonded abrasive machining when the free abrasives are captured and held by the tool. Consequently, machining of the material through shear can take place, resulting in visible scratches on the surface. The photo in Figure 3.8, with a wafer surface that carries long scratches resulting from a FAM experiment, demonstrates an example of the scratching and shearing mechanism, as well as rolling-indenting and rolling-scratching which results in randomly-distributed surface pits, as those in Figure 3.7.

The lateral crack from indentation fracture, shown earlier in Figure 3.2(b), is a special characteristic of brittle materials [Lawn and Evans (1980)] and is considered as the main contributor to brittle material removal. However, the control of

Figure 3.8 Long scratches resulting from the lapping experiment with 9 μm mean size SiC abrasives and 5 lb loading on a 3 in silicon wafer. Source: Imin Kao.

surface roughness of wafers subject to such machining processes is not as precise as that of ductile machining. In addition, the accompanying median cracks and the lateral cracks beneath the wafer surface, called the subsurface damage (SSD), have to be removed in the subsequent processing steps after slicing. There is a threshold to reach the indentation fracture. It depends on the material properties and the depth of penetration. Therefore, it is possible for the brittle material to be machined like ductile material, as presented in Section 3.6.

Although most research has treated the lateral crack as a by-product of the free abrasive machining process, the contribution of the ductile-regime machining may not be entirely neglected, as discussed in Section 3.6. Therefore, there can be up to a total of four machining mechanisms to explain the complex free abrasive machining process, including both the brittle and ductile modes of machining.

3.7.2 Comparison Between Wiresawing and Lapping

From the previous sections, two free abrasive machining processes, wiresawing and lapping, are described. They are employed for different purposes in wafer manufacturing. A modern slurry wiresaw is used for slicing, while lapping is for removing the surface tool marks and subsurface damage resulting from slicing. In this section, the two processes will be compared.

Figure 3.9 illustrates the involvement of different machining mechanisms of free abrasive machining for wiresawing, lapping, and polishing [Bhagavat et al. (2005, 2010)]. In the wiresawing process, the rolling-indenting mechanism dominates the machining process because the hydrodynamic effect enlarges the gap between the workpiece and the high-speed wire. By contrast, the abrasives in lapping process, with smaller relative motion between the tool and workpiece, will make direct contact with both workpiece and tool, causing both rolling-indenting and

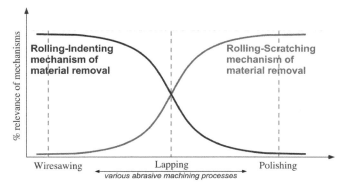

Figure 3.9 Models of the two mechanisms of FAM processes: rolling-indenting and rolling-scratching, as plotted against various FAM processes with percentage of relevance. Wiresawing primarily involves rolling-indenting. Both rolling-indenting and rolling-scratching are involved in lapping depending on process parameters. Polishing primarily involves very gentle rolling-scratching.

Table 3.4 Comparison between slicing using a slurry wiresaw and lapping in wafer manufacturing.

	Wiresaw slicing	Lapping
Purpose	Forming and slicing	Flatten surface; remove subsurface damage
Machining area	Smaller (underneath a wire)	Larger (entire wafer surface)
Material removed	High	Low
Machining direction	Tangent to the wafer surface	Normal to the wafer surface
Machining process	Rolling-indenting	Rolling-indenting, rolling-scratching, and scratching-shearing
Abrasives	Silicon carbide (larger)	Alumina oxide (smaller)
Surface topography	Warp, waviness, indentation, and fracture	Indentation, scratch, and fracture

rolling-scratching mechanisms to happen during the machining process. Table 3.4 compares these two machining tools.

3.7.3 Other Aspects of Engineering Modeling

Elastohydrodynamic Process Modeling
When abrasion and relative motion between surfaces or tools take place with fluid in the contact interface, the hydrodynamic effect must be considered. Even a conventional wire drawing process with light lubrication at moderate speed will incur the hydrodynamic effect. Hydrodynamics in the presence of elasticity, such as the wire in the wiresawing process, will result in elastohydrodynamic interaction. The process modeling is complex with engineering multi-physics modeling and analysis. This subject will be discussed in further details in Chapter 5.

Vibration Analysis
The continuum tools used in the BAM and FAM processes, such as wire in the wiresaw or lapping plate in lapping, will display vibration behavior to different extents during the machining process, depending on the speed and support (boundary conditions) of the continuum tools. Vibration has profound implications on the quality of surface and subsurface damage in the BAM and FAM processes. Specifically, the analysis of vibration and responses of a moving wire under high tension is an important consideration in a wiresawing process. Although vibration of a stationary wire is a century-old subject with well-known solutions, the eigenvalue problem of vibration of a moving wire only recently became a topic of research, both for free vibration and forced vibration. This subject will also be discussed further in Chapter 5.

3.8   Abrasive Slurry in FAM Processes

In this section, the composition of abrasive slurry commonly used in FAM processes are discussed. Slurry used by wiresaws serves very important purposes. In lapping, the slurry composition is different from that of wiresawing but shares the same general ingredients.

3.8.1   Composition of Abrasive Slurry

An abrasive slurry is crucial in wafer slicing using the modern slurry wiresaws. Abrasive slurry employed in the wiresaw slicing process consists of abrasive grits and carrier fluid, as described in the following.

- Abrasive grits: This refers to the abrasive material in fine grains that are to be suspended by the carrier fluid for performing the machining to remove materials from the contact interface. The abrasive material has to be much harder than the material to be machined. As an example, silicon carbide is often used as an abrasive material to slice silicon ingots into wafers. As a comparison, when silicon carbide is used to slice aluminum oxide ingots, it will take at least 10 times longer than slicing a silicon ingot of the same size (cf. Table 3.3 for the comparison of hardness). Thus, the selection of abrasive material is critical in slicing. Common abrasive materials used in wafer manufacturing include silicon carbide, boron carbide, and industrial diamond, in order of increasing cost.
- Carrier fluid: The abrasive grits are mixed with carrier fluid in order that the abrasive particles are suspended in the fluid for machining. This facilitates the three-body abrasion in FAM, as discussed earlier. Typical carrier fluid includes oil, glycol, and water. Oil has high viscosity and consistency to suspend the abrasive grains well. However, oil as a carrier fluid has two major problems: oil breakdown and environmental concerns. The oil-based slurry is not easy to dispose of because it is not water soluble and may cause environmental concerns in land fill. Nowadays, more operations in wafer production have turned to water-soluble or water-based carriers. Glycol, being water soluble and more viscous, is a common carrier fluid in wiresaws. When water is used, typically in a lapping process, de-ionized (DI) water is used.

Abrasive grits and carrier fluid are mixed with a prescribed ratio between the mass of the grits and the volume of the carrier (in liters). We define this slurry mixing ratio as

$$r_s = \frac{m_g}{V} \tag{3.1}$$

where m_g is the mass of the abrasive grits in kilograms, and V is the volume of the carrier fluid in liters. For example with one kilogram of silicon carbide and one liter of glycol, the slurry mixing ratio will be $r_s = 1.0$ kg L^{-1}. Naturally, the viscosity of the slurry is related to the ratio between the abrasive grits and carrier. A chart for different mixing ratios ($r_s = 0.75 \sim 1.25$ kg L^{-1}) of SiC in water-soluble carrier is shown in Figure 3.10. In the figure, the viscosity (in centipoise, *cP*) is plotted in logarithmic

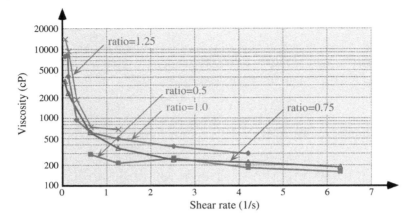

Figure 3.10 Experimental results on the relationship of the viscosity and shear rate for a water-soluble carrier with SiC. The four different mixing ratios represent a kilogram of SiC abrasive grit per liter of carrier. Source: Kao (2004).

scale with respect to the shear rate. Four different slurry mixing ratios are plotted. The mixing ratio of $r_s = 1.25$ results in the maximum viscosity of 14000cP at low shear rate. It is clear from Figure 3.10 that higher ratios result in higher viscosities. On the other hand, if the ratio is too low the slurry will lose its viscosity, resulting in poor cut. A commonly accepted mixing ratio is between $r_s = 0.75$ and 1.0.

When mixed well, the slurry of a wiresaw operation is pumped into the nozzle, positioned on top of a wire web (cf. Chapter 4) for slicing. Abrasive grits account for the majority of the cost in a slurry for wiresawing operation. In a lapping operation, DI water is typically mix with aluminum oxide grits and dripped slowly into the contact interface of lapping. The slurry mixing ratio of lapping operation tend to be smaller than the wiresawing operation.

3.8.2 Comparison of Water and Glycol as a Carrier Fluid for Slurry

The size of abrasive grits and properties of carrier fluid will affect the performance of the cutting process and parameters such as viscosity of the slurry. In addition, the slurry mixing ratio, defined in Equation (3.1), also has significant influence on the viscosity of the slurry. These properties will in turn affect the hydrodynamic and vibration characteristics of the modeling. The common carriers employed in the wiresaw system are either oil-based or water-soluble (including water-based). The water content in water-soluble (e.g. glycol) carrier can affect the cutting performance. In this section, water-soluble carrier fluid is employed to mix slurry with silicon carbide abrasives in the wiresawing process to investigate the impact on the surface properties of the sliced wafers. The experiments were conducted by using glycol, a water-soluble carrier fluid, with different water content.

Figure 3.11 shows empirical data on the water content, at 0% and 25%, in water-soluble slurry versus cutting efficiency [Kao et al. (1998b)]. The results

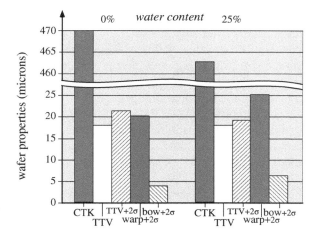

Figure 3.11 Experimental results on the effect of slurry water content on the performance of slicing 150 mm silicon wafers. The experiments were conducted with 0% and 25% water content. The symbols are as follows: CTK – center thickness, TTV – total thickness variation, warp – measure of surface unevenness, bow of wafer – a measure of the waviness of the surface. Source: Kao et al. (1998b) © The Electrochemical Society. Reproduced by permission of IOP Publishing. All rights reserved.

are obtained from cutting hundreds of wafers. The surfaces of the wafers are characterized by TTV, warp and bow, introduced in Section 1.4.1, along with the thickness at the center (CTK). From the figure, it is found that the center thickness (CTK) is greatly reduced when the slurry contains 25% water. Although TTV remains the same, the warp and bow are both higher for wafers cut with slurry of 25% water content. The standard deviation of TTV for slurry with 0% water content is slightly higher but the difference is very small (a couple of micrometers for 2σ distribution). Therefore, the results of bow and warp suggest that the inclusion of water in water-soluble slurry tends to degrade the performance of the slurry in the wafer slicing process. This is a concern because the moisture in the ambient will be absorbed by the carrier fluid, such as glycol. Due to the degradation, a regeneration process that separates the grits from carrier after the slurry has performed operations for a certain period of time is desirable. The separation of slurry from carrier fluid and recycling of abrasive grits will be discussed in the following section. The grits can be recycled and mixed with a new carrier for slicing.

3.8.3 Recycling of Abrasive Grits in Slurry

When the slurry is aggregated with the silicon kerf (the small grains of silicon removed from the ingot during slicing process), it becomes more difficult for the slurry to perform the slicing function due to the contamination of silicon kerf in abrasives slurry. Thus, a slurry is reused no more than five times when slicing with a wiresaw in order to keep efficiency of slicing and to maintain the quality of the sliced wafer surface. That is, each new batch of slurry is used to slice five ingots. In the past, the spent slurry was discarded along with silicon kerf that was generated with cutting to landfill. In order to reduce the cost of abrasive materials, and to maintain the efficiency in producing silicon wafers with nice surface quality, the recovery of silicon carbide from the spent slurry becomes increasingly important. The recovered silicon carbide can be mixed with new carrier fluid and be used for slicing, to reduce the cost. In addition, discarding the spent slurry will create environmental concerns. In an age of emphasis on sustainability, proper handling of the spent slurry has developed into a new opportunity. Specifically, the recovery of abrasive grits or silicon kerf has provided a new market of abrasive slurry recycling [Costantini et al. (2000); Fallavollita (2011); Kozasa and Gotou (2009); Lan et al. (2011); Watson and Glaser (1966)].

The most common method of recovery of abrasives from spent slurry is to use a centrifuge to separate silicon carbide from the silicon kerf. The efficiency of the recycling process and detection of the quality of abrasive slurry has developed a new market of tools and instruments. Laser diffraction and imaging methods engaged with a CCD camera were used to help implement recycling process monitoring. This can help detect deterioration of abrasive grits and slurries in order to determine when a slurry should be recycled. In addition to the deterioration of abrasive grits, other challenges include the yield of silicon carbide recovery, reduction of cost of recycling, potency of recycled abrasives, and environmental issues relevant to spent slurry and recovery processes.

Figure 3.12 compares two SEM photos of F-400 SiC abrasive grits; one is fresh SiC abrasive grits, the other is recovered SiC from spent slurry used in wiresaw slicing. The recycled SiC abrasives show silicon kerfs interspersed in the photo of SiC abrasives. In some experiments, it was shown that the regenerated abrasive slurry, which inevitably will include a very small amount of silicon kerf, can enhance the slicing performance. Under such situation when the recycled abrasives are mixed with brand new abrasives, the new slurry with mixed recycled abrasives works better than

(a) fresh SiC   (b) recovered SiC

Figure 3.12 SEM photos of (a) fresh SiC abrasive grits (F-400 grits), and (b) recovered SiC from spent slurry used in wiresaw slicing, showing the silicon kerf interspersed in the photo of SiC abrasives (F-400). Source: Imin Kao.

totally brand new abrasives, and produces wafers with smaller TTV and warp. Such observation is consistent with the finding that a fresh batch of slurry with brand new abrasives tends to produce wafers with better surface properties (such as warp, bow, and TTV) when slicing the second or third ingots, after slicing the first ingot. This is due to the small amount of silicon kerf existing in the slurry along with the SiC abrasive grits.

3.9   Summary

In this chapter, the modeling of ductile and brittle manufacturing processes, in the context of wafer manufacturing, is presented with attention given to the fundamental differences between the ductile and brittle machining processes. Recent research on ductile machining of brittle materials was presented with literature references, as well as the potential in the future advance of wafer manufacturing using the ductile mode of machining on semiconductor materials. This is followed by the abrasive machining pertaining to wafer manufacturing, which can be broken down into two categories: (i) bonded abrasive machining (BAM), and (ii) free abrasive machining (FAM). Processes of wafer manufacturing are discussed with respect to the two categories of abrasive machining. The classification of grain size and hardness characterization of abrasive materials are presented. Abrasive materials that are commonly used in wafer manufacturing processes are described in detail. The three process models of abrasive machining, including (i) rolling-indenting, (ii) scratching-indenting, and (iii) scratching-shearing models, are presented with examples. Different surface morphologies of the machining models were illustrated to elucidate the process of abrasive machining. Various experiments to investigate the properties of slurry in slicing using a wiresaw were presented with summary of results. References of literature are included for further reading in this area of research.

References

Ajjarapu SK, Fesperman RR, Patten JA and Cherukuri HP 2004 Ductile regime machining of silicon nitride: A numerical study using Drucker-Prager material model. *NAMRC* **32**, 519.

ANSI 2006 American National Standard ANSI-B74.18-2006 For Grading of Certain Abrasive Grain on – Coated Abrasive Material, sponsored by the United Abrasives Manufacturers' Association (UAMA). .

ANSI 2009 American National Standard ANSI-B74.12-2009, Revision of B74.12-2001 Specification for the Size of Abrasive Grain – Grinding Wheels, Polishing and General Uses. URL http://www.superabrasives.org/pdfs/B74_12_2009.pdf.

ANSI 2010 American National Standard ANSI-B74.10-2010, Revision of B74.10-2001 Specification for – Grading of Abrasive Microgrits.

Bhagavat M, Prasad V and Kao I 2000 Elasto-hydrodynamic interaction in the free abrasive wafer slicing using a wiresaw: Modeling and finite element analysis. *Journal of Tribology* **122**(2), 394–404.

Bhagavat S, Liberato J and Kao I 2005 Effects of mixed abrasive slurries on free abrasive machining processes *Proceedings of the 2005 ASPE Conference* ASPE.

Bhagavat S, Liberato J, Chung C and Kao I 2010 Effects of mixed abrasive grits in slurries on free abrasive machining (FAM) processes. *International Journal of Machine Tools and Manufacture* **50**, 843–847.

Blackley W and Scattergood RO 1991 Ductile-regime machining model for diamond turning of brittle materials. *Precision Engineering* **13**(2), 95–103.

Blake PN and Scattergood R 1990 Ductile-regime machining of germanium and silicon. *J. Am. Ceram. Soc.* **73**(4), 949–957.

Callister WD and Rethwisch DG 2014 *Materials Science and Engineering: An Introduction.* John Wiley and Sons, Inc., New Jersey.

Capraro B 2013 The move to the next silicon wafer size *SEMICON, Europa 2013* European 450mm Equipment & Materials Initiative: EEMI 450.

Chang YP, Hashimura M and Dornfeld DA 2000 An investigation of material removal mechanisms in lapping with grain size transition. *Journal of Manufacturing Science and Engineering* **122**, 413–419.

Chonan S and Hayase T 1987 Stress analysis of a spinning annular disk to a stationary distributed, in-plane edge load. *J. of Vibration Acoustics Stress and Reliability in Design* **107**, 277–282.

Chonan S, Jiang ZW and Yuki Y 1993 Stress analysis of a silicon-wafer slicer cutting the crystal ingot. *Journal of Mechanical Design* **115**, 711–717.

Chung C and Kao I 2008 Comparison of free abrasive machining processes in wafer manufacturing *Proceedings of International Manufacturing Science and Engineering Conference (MSEC 2008)*, vol. ASME Paper number MSEC2008-72253 ASME.

Clark W, Shih A, Hardin C, Lemaster R and McSpadden S 2003a Fixed abrasive diamond wire machining- part i: Process monitoring and wire tension force. *International Journal of Machine Tools and Manufacture* **43**(5), 523–532.

Clark W, Shih A, Lemaster R and McSpadden S 2003b Fixed abrasive diamond wire machining- part ii: Experiments design and results. *International Journal of Machine Tools and Manufacture* **43**(5), 533–542.

Costantini M, Talbott J, Chandra M, Prasad V, Caster A, Gupta K and Leyvraz P 2000 Method and apparatus for improved wire saw slurry. US Patent 6,113,473.

Fallavollita JA 2011 Methods and apparatus for recovery of silicon and silicon carbide from spent wafer-sawing slurry. US Patent App. 12/865,989.

FEPA 2014 The abrasives website of the federation of the European producers of abrasives Website. URL http://www.fepa-abrasives.org/.

Fine Tools 2014 Conversion Chart Abrasives – Grit Sizes, from Dieter Schmid - Fine Tools Website. URL http://www.fine-tools.com/G10019.htm.

Fuchs BA, Hed PP and Baker P 1986 Fine diamond turning. of KDP crystals. *Applied Optics* **25**, 1733–1735.

Groover MP 2012 *Fundamentals of Modern Manufacturing: Materials, Processes, and Systems* 5th edn. Wiley.

Intel News Release 2008 Intel, Samsung Electronics, TSMC reach agreement for 450mm wafer manufacturing transition Website. URL http://www.intel.com/pressroom/archive/releases/2008/20080505corp.htm.

Kakani SL and Kakani A 2004 *Material Science*. New Age International Publishers.

Kalpakjian S and Schmid S 2013 *Manufacturing Engineering & Technology* 7th edn. Prentice Hall.

Kao I 2004 Technology and research of slurry wiresaw manufacturing systems in wafer slicing with free abrasive machining. *the International Journal of Advanced Manufacturing Systems* **7**(2), 7–20.

Kao I and Bhagavat S 2007 Single-crystalline silicon wafer production using wire saw for wafer slicing Transworld Research Network Editors J. Yan and J. Patten, Kerala, India chapter 7, pp. 243–270.

Kao I, Bhagavat M and Prasad V 1998a Integrated modeling of wiresaw in wafer slicing *NSF Design and Manufacturing Grantees Conference*, pp. 425–426, Monterey, Mexico.

Kao I, Chung C and Rodriguez R 2010 *Modern Wafer Manufacturing and Slicing of Crystalline Ingots to Wafers Using Modern Wiresaw* Springer Handbook of Crystal Growth; *ed.* G. Dhanaraj M. Dudley, K. Byrappa, and V. Prasad Springer-Verlag chapter 52.

Kao I, Prasad V, Chiang FP, Bhagavat M, Wei S, Chandra M, Costantini M, Leyvraz P, Talbott J and Gupta K 1998b Modeling and experiments on wiresaw for large silicon wafer manufacturing, *The 8th Int. Symp. on Silicon Mat. Sci. and Tech.*, p. 320, San Diego.

Klein C 2008 *Minerals and Rocks: Exercises in Crystal and Mineral Chemistry, Crystallography, X-ray Powder Diffraction, Mineral and Rock Identification, and Ore Mineralogy* 3rd edn. John Wiley & Sons.

Kojima M, Tomizawa A and Takase J 1990 Development of new wafer slicing equipment (unidirectional multi wire-saw). *Sumitomo Metals* **42**(4), 218–224.

Kozasa K and Gotou I 2009 Method of recycling abrasive slurry. US Patent App. 12/192,351.

Lan C, Lin Y, Wang T and Tai Y 2011 Recovery method of silicon slurry. US Patent 8,034,313.

Lawn BR 1993 *Fracture of Brittle Solids* 2nd edn. Cambridge Univ. Press.

Lawn BR and Evans AG 1980 Elastic/plastic indentation damage in ceramics: the median/radial crack system. *J. Amer. Ceram. Soc.* **63**, 574–581.

Lawn BR and Swain MV 1975 Microfracture beneath point indentations in brittle solids. *J. Mater. Sci.* **10**, 113–122.

le Scanff A 1988 New wire saw machine. *Industrial Diamond Review.*

Lely JA 1955 Darstellung von einkristallen von silicium carbid und beherrschung von art und menge der eingebauten verunreinigungen. *Berichte der Deutschen Keramischen Gesellschaft* **32**, 229–236.

Li J, Kao I and Prasad V 1998 Modeling stresses of contacts in wiresaw slicing of polycrystalline and crystalline ingots: Application to silicon wafer production. *Journal of Electronic Packaging* **120**(2), 123–128.

Mech HW 1974 Machine and method for cutting brittle materials using a reciprocating cutting wire. *U.S. Patent.*

Morris JC, Callahan DL, Kulik J, Patten JA and Scattergood RO 1995 Origins of the ductile regime in single-point diamond turning of semiconductors. *J. Am. Ceram. Soc.* **78**(8), 2015–20.

Ngoi BKA and Sreejith PS 2000 Ductile regime finish machining – a review. *International Journal of Advanced Manufacturing Technology* **16**, 547–550.

Patten J, Cherukuri H and Yan J 2004 *Ductile regime machining of semiconductors and ceramics.* vol. High-Pressure Surface Science and Engineering. CRC Press. Book Chapter in High Pressure Surface Science and Engineering edited by Y. Gogotsi and Y. Domnich.

Patten J, Gao W and Yasuto K 2005 Ductile regime nanomachining of single-crystal silicon carbide. *Journal of Manufacturing Science and Engineering, Transactions of the ASME* **127**, 522–532.

Patten JA and Gao W 2001 Extreme negative rake angle technique for single point diamond nano-cutting of silicon. *Precision Engineering* **25**, 165–167.

Puttick KE, Rudman MR, Smith KJ, Franks A and Lindsey K 1989 Single-point diamond machining of glasses. *Proc. Roy. Soc.* **A.426**, 19–30.

Sahoo R, Prasad V, Kao I, Talbott J and Gupta K 1998 Towards an integrated approach for analysis and design of wafer slicing by a wire saw. *Journal of Electronic Packaging* **120**(1), 35–40.

Shibata T, Fuji S, Makino E and Ikeda M 1996 Ductile-regime turning mechanism of single-crystal silicon. *Precision Engineering* **18**(2/3), 129–137.

Shimizu H 1976 Wire-saw. *U.S. Patent.*

Smokintbird 2009 Stone, Belt, Paper, Film and Compound Grit Comparison – Downloadable PDF file compiled by Smokintbird, with help of Olivia on SRP forum. URL http://www.google.com/F#q=compiled+by+Smokintbird.

UAMA 2014 The Unified Abrasives Manufacturers' Association website. URL http://www.uama.org/index.html.

Washington Mills 2014 Washington Mills website Website. URLs http://www.washingtonmills.no/products/grit-sizes.html; http://www.washingtonmills.com/.

Watanabe M and Kramer S 2006 450 mm silicon: An opportunity and wafer scaling. *The Electrochemical Society Interface* **15**(4), 28–31.

Watson WI and Glaser PW 1966 Silicon carbide recovery. US Patent 3,259,243.

Wells R 1987 Wire saw slicing of large diameter crystals. *Solid State Technology* pp. 63–65.

Wells RC 1985 Wire saw. *U.S. Patent.*

Wikipedia 2014a Diamond –From Wikipedia, the free encyclopedia. URL http://en.wikipedia.org/wiki/Diamond.

Wikipedia 2014b Sic –From Wikipedia, the free encyclopedia. URL http://en.wikipedia.org/wiki/Silicon_carbide.

Part II

Wafer Forming

4

Wafer Slicing Using a Modern Slurry Wiresaw and Other Saws

4.1   Introduction

This chapter begins a series of presentations on *wafer forming* after crystal growth. The first post-growth wafer forming process is the slicing of ingot into wafers. Today, wafer slicing is performed by various types of saws, including, but not limited to, inner-diameter (ID) saws, modern slurry wiresaws, and diamond-impregnated wiresaws. In fact, the majority of current wafer slicing is performed using wiresaws, which will thus be the focus of this chapter.

A saw that utilizes wire as a machining tool is not a new technology. Such saws were used beginning in 19th century Europe on slab stones; the carpenter's wiresaw utilized a steel wire with notches as a cutting tool to carve through wood workpieces. The modern slurry wiresaw, in contrast, is quite different from wiresaws used centuries ago. The modern wiresaw is employed to produce semiconductor prime wafers and photovoltaic (PV) wafers, which require more rigorous processes and stringent requirements of the surface quality (a few microns of surface variation, or less). Modern microelectronics fabrication became a leading industry in the second half of the 20th century, and products using microelectronic chips create many consumer electronics in the 21st century, including PCs, laptops, tablets, and smart phones. Prime wafers, a product of wafer manufacturing, as discussed in Chapter 2, are an essential part of modern microelectronics fabrication. The process of wafer manufacturing and various growth processes of crystalline ingots were introduced in Chapter 2. With continuously improving processes and practices in industry, it is possible to make larger and better ingots with different materials. Slicing is the first post-growth process for wafer forming. The modern slurry wiresaw was first designed and introduced to the industry for PV wafering in the early 1990s. It gained widespread adoption in the mainstream microelectronics industry where it has been employed to produce single crystalline silicon wafers since the late 1990s. Furthermore, it has replaced nearly all ID saws for silicon wafer production. Such a wholesale change in machine tools is highly dependent upon the large throughput of wafer slicing, quality of wafers produced, the ability to slice wafers of large diameters, and the economy of wafer production.

Wafer Manufacturing: Shaping of Single Crystal Silicon Wafers,
First Edition. Imin Kao and Chunhui Chung.
© 2021 John Wiley & Sons Ltd. Published 2021 by John Wiley & Sons Ltd.

In this chapter, the modern slurry wiresaw as a machine tool for slicing is introduced and compared with ID saws and other saws.

4.2 The Modern Wiresaw Technology

Before moving on to the presentation of wafer forming in this chapter and this book, it is of critical importance to properly define the terminology for **wiresaw** as it is used throughout this book.

As discussed earlier in Section 2.4.6, the term *wiresaw* used in this book specifically refers to a class of machine tools that utilize a single wire to span a wire web surface for slicing crystal ingots, with abrasive slurry. In this book, the terms *wiresaw* and *slurry wiresaw* are used interchangeably. Many different names can be found in the literature for this manufacturing machine tool. Some called it a "multi-wire saw", which is misleading because only one continuous wire is used to span a wire web that appears to be composed of multiple wires. This is in contrast to the multiple wires used in a configuration of a diamond saw with multiple segments of diamond-impregnated wires. Some others simply call it "wire saw," but this term can be a reference to many different types of machine tools using a wire or wires; thus, it lacks clarity.

Therefore, in order to maintain the specificity of the reference to this specialized machine tool using a wire web, formed by a single wire, immersed in abrasive slurry for wafering, we shall call such a machine tool "wiresaw" throughout this book.

4.2.1 Historical Perspectives of Saws Using Wire

Wires have been employed as a machining tool since the 19th century in Europe to slab stones. Figure 4.1 shows a photo depicting the use of a wire as a new method of slate-mining, without using dynamites, in 1901 by the Welsh mining industry. According to an article published in Nature (1901), "At Labassère the wire saw is employed to make horizontal cuts across the inclined beds of slate, severing off great blocks without blasting."

This article demonstrated the installation of a helicoidal wire. The wire, made of three hard wires twisted together to form "an endless cord", traveled along the cutting interface by machinery and pulleys, though time to slice rocks was not reported in the article. It had an uneven surface for machining, due to the formation of the twisted triplet wires, and was fed continuously with sharp particles of sand and water. A schematic of the wire saw at the contact interface between the wire and the rock is illustrated in Figure 4.2(a). A magnified view shows the sharp particles of sand and water forming a slurry (cf. Section 3.8) to perform free abrasive machining (discussed in Section 3.4) with the wire as the tool that carries the slurry and free abrasives.

Another wire-based saw is the carpenter's wire saw, as illustrated in Figure 4.2(b). Here, a bow is used to stretch and apply tension to a steel wire, where the tension prevents the wire from premature breakage due to low stiffness and bending with

Figure 4.1 A photo of a wire for quarrying in 1901, a new method of slate mining, using a wire saw. Source: Nature (1901).

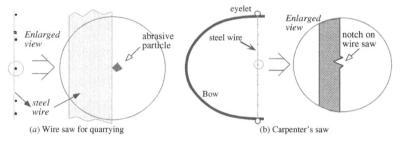

(a) Wire saw for quarrying   (b) Carpenter's saw

Figure 4.2 (a) A schematic of the wire saw for quarrying in 1901 (free abrasive machining, FAM); (b) An illustration of a carpenter's wire saw (bonded abrasive machining, BAM).

stress concentration at the notches. The wire is also sparsely cut with notches on the surface along the longitudinal direction, as illustrated in Figure 4.2(b). When a carpenter uses this type of saw, he moves the saw in a reciprocating motion to produce the cutting operation on a wood workpiece, equivalent to a conventional saw in woodworking. In the context of abrasive machining, a carpenter's wire saw can be regarded as akin to the bonded abrasive machining (BAM) process described in Section 3.4, with the metal notches as bonded abrasive grits, in a ploughing process to remove material from the workpiece.

This is in contrast to the triple-twisted wire for quarrying in Figure 4.2(a), in which the wire moving along the surface of the stone is accompanied by water and free abrasive (sand) particles. The abrasive particles that are caught between the wire and contact surface are, to a certain extent, equivalent to the notches of a carpenter's wire saw at the instant of machining. This action, combined with the application of the subsurface maximum shear stress due to contact, as well as surface and subsurface fractures, produces the cutting action. For the quarrying saw, the reciprocating motions help the cut portion to be carried away and agitate abrasive particles to join the cutting operations.

Some common issues are observed in machining when using a carpenter's wire saw. For example, the saw does not always produce a straight cut because the manually operated saw has very small diameter and, without proper control, can easily wander from its prescribed path on a soft material like wood. In addition, it is important to apply appropriate tension and stiffness to maintain a straight cut.

In the late 20th century, wire saws with carborundum (SiC) abrasive slurry were used in the slabbing of granite blocks (especially those over 1.7 m in height) [le Scanff (1988)]. The stone cutting industry also used saws with wire made by diamond-impregnated beads [Contardi (1993)] which provided a low-vibration solution to concrete cutting [Hayes (1990); Herbert (1989)].

4.2.2 The Rise of the PV Industry and Wafer Slicing

From 1972 to 1992, the cost of photovoltaic (PV) modules was reduced 100-fold, spurring an increase in the size of the global silicon PV manufacturing market to 91.3 billion USD in 2011, 80 billion USD in 2012, and 91.3 billion USD in 2013 [Statista (2014)]. This growth in market size was made possible by a reduction in cost at every stage of PV manufacturing, although wiresaws have played a very important role in the substantial decrease in wafer thickness and kerf loss. Starting in the early 1990s, the PV industry has greatly benefited from the introduction of wiresaw technology [Kojima et al. (1990); Mitchell et al. (1994)]. Modern slurry wiresaws provide an excellent machining technology for PV wafer production because of their ability to slice both single crystalline and polycrystalline silicon ingots of different sizes and shapes into very thin wafers with a small kerf loss[1] and high yield. Since the mid 1990s, modern slurry wiresaws have been commonly used for the production of PV silicon wafers at thicknesses of 200–400 µm with very small kerf losses of 150–230 µm. During this period, the development of wiresaw technology and machinery has largely taken place in Europe and Japan.

After the mid of 1990s, the microelectronics industry started to take an interest in wafer production using modern slurry wiresaws. However, the requirements for wafer slicing in microelectronics and PV applications are much more stringent and rigorous than those of stone quarrying, with demand for very good surface quality. Most silicon PV wafers are processed to produce solar cells as-sliced without surface lapping or polishing. Such applications require good surface finish after slicing. The microelectronics industry, on the other hand, requires prime wafers with good surface quality after slicing, before proceeding with the lapping, grinding, and polishing processes. Furthermore, with the increase of the diameter of silicon wafers to 300 mm, the modern slurry wiresaw has become the only viable technology for slicing. Wiresaws can also be employed to cut harder electronic materials such as

---

1 The *kerf loss* refers to the materials that are removed by the saw during the cutting process.

certain III–V compounds (e.g., GaAs and InP) or ceramic materials (e.g., silicon carbide and quartz), given that the process is accurately modeled, understood, and controlled.

4.3 The Three Categories of Saw for Wafer Slicing

Different saws designed for wafer slicing (since the 1960s) include the following three main machine tools:

- The inner diameter (ID) saw.
- The modern slurry wiresaw or *wiresaw*.
- Saws with moving wires impregnated by or coated with diamond abrasives.

The ID saw will be presented in Section 4.4. The modern slurry wiresaw (or wiresaw) will be further discussed in Section 4.5 in more detail. A comparison between the wiresaw and ID saw is offered in Section 4.6.

The saws that utilize wires with impregnated diamond abrasive grits will be presented in Chapter 6 of this book. However, it is important to note that the modern slurry wiresaw (cf. Chapter 3), in terms of the mechanism of machining, is fundamentally different from the diamond-impregnated wiresaw. This difference in the machining mechanisms is presented and discussed in Section 3.4. A diamond wiresaw uses wires coated with diamond grits on the circumferential surface of the wire to remove material [Clark et al. (2003a,b)]. The saws with diamond-impregnated wire are essentially an ID saw with a linear cutting edge instead of an annular one. One of the serious weaknesses of the diamond impregnated wiresaw is the stripping of diamond grains from the peripheral surface of the wire due to the high shear generated in the cutting process. Unlike the abrasive slurry wiresaw where the "free" abrasives will re-join the cutting process, the diamond coated wire will lose its cutting capability once the bonding between the surface of the wire and the diamond grains is lost. In addition, the equipment and consumables of diamond impregnated wires can be more expensive per wafer, making it challenging for the current technology to slice large-diameter silicon wafers in high volume production without technological improvements. As the technology of materials and bonding improve, this technology will become more economical and competitive as a tool for wafer slicing. Saws with diamond-impregnated wire can also be an effective machine tool for small-volume production of small crystal boules and for customized operations in specialized slicing processes.

4.4 Inner-diameter (ID) Saw

The inner diameter saw (ID saw) is a machine tool designed for slicing wafers from ingots. The conventional ID saw cuts through ingots to obtain one slice of wafer each time, with an annular cutting blade stretched and held at very high tension. The blade rotates at very high angular speed, ω, while feeding onto the ingot, and

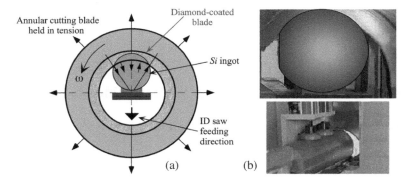

Figure 4.3 (a) Schematic of a conventional ID saw for wafer slicing; (b) the front (top) and back (bottom) views of a ID saw with ingot mounted for slicing and the ID blade pressing on the ingot, showing fixtures holding the ingo. Source: Imin Kao.

cuts through the ingot with its diamond-coated ring along its inner peripheral edge. A schematic of an ID saw is illustrated in Figure 4.3(a). The ingot is mounted on a fixture, as shown, that fits in the inner hole of the ID saw. As the ID saw traverses downward, the bonded abrasives on the cutting edge of the ID blade slice through the ingot to produce one slice of wafer at a time.

Figure 4.3(b) shows a photo of the front view of an ingot being sliced by an ID saw, and the fixture that holds the ingot in the rear view. The ID saw has been used for slicing silicon ingots since the early days of silicon wafer production for microelectronics fabrication. The largest diameter of ingots that can be effectively sliced by ID saws is 200 mm, or 8 in. Slicing wafers of 200 mm diameter is already pushing the limit of the capability of ID saw technology. It was reported that an ID saw for slicing 300 mm wafers was designed, but failed to gain traction due to the advent of wiresaws. As the diameter of silicon wafers increases, ID saws become less effective and more dispensable, due to the longer average time it requires to produce one wafer. In an ID saw manufacturing plant for wafer production, one sees several rows of many ID saws operating at the same time in order to produce the throughput required in wafer production as each ID saw can only produce one wafer at a time.

SEM photos of the surface of an as-sliced silicon wafer by an ID saw are shown in Figure 4.4. In the left photo, deep grooves of uneven tool marks are seen clearly with visible surface fractures. In a zoom-in view of the surface on the right, fractures and damage on the surface can be seen easily due to the ploughing process of machining by an ID saw. In addition, the path of diamond abrasive grits ploughing through the surface of the wafer can be seen with different depths of cut on the surface.

Diamond-coated edges of an ID saw become less effective after slicing through many pieces of silicon wafer. This can be attributed to (1) wear of the cutting edge of diamond grains and (2) the silicon particles left on the cutting surface, which need to be regenerated in a truing process using a dressing stick, as shown in Figure 4.5. The dressing sticks are typically made of green silicon carbide. In a truing process, a silicon carbide dressing stick is fed through, and sliced by the ID saw, in order to

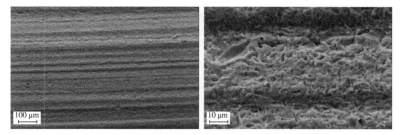

Figure 4.4   SEM photos of the surface sliced by an ID saw: (left) photo with visible tool marks with deep grooves and surface fractures; scale of 100 μm is shown, (right) fractures on the surface of the wafer; scale of 10 μm is shown. Source: Imin Kao.

Figure 4.5   A photo of a dressing stick made of green silicon carbide, used in reconditioning and regenerating the diamond cutting surface of ID saws in a truing process. Source: Imin Kao.

regenerate the cutting surface with the diamond abrasive grits and their cutting edge.

The vibration of the ID blade will determine, along with the thickness of the steel blade and the diamond coating, the kerf loss. Hence, vibration analysis and the stress analysis are important in understanding and improving the design and operation of ID saws. Study and analysis of stress and vibration of an ID saw with annular blades were presented in [Chonan and Hayase (1987); Chonan et al. (1993a,b)].

4.5   The Modern Slurry Wiresaw

The modern *wiresaws* utilize a wire web spun from a single steel wire immersed in abrasive slurry, consisting of abrasive grits suspended in carrier fluid, to slice through crystalline ingots to produce wafers. A photo of a HCT wiresaw is shown in Figure 4.6, with many competing companies marketing and selling wiresaws of different designs. Nevertheless, a typical wiresaw consists of the following three main components:

(1) The control and program console
(2) The wire management unit
(3) The slicing compartment upon which the to-be-sliced ingot sits.

Figure 4.6 Photo of a modern industrial wiresaw for wafer slicing. Source: GT Equipment Inc.

4.5.1 The Control and Program Console

This console consists of display and interactive panels, as well as input devices for operators to configure the system and set up parameters of the machining operation. The process parameters are programmed for automated slicing, including the speed of the wire, the feed rate of the ingot onto the wire web, the length of wire in either direction of reciprocating motion in slicing, slurry management, and the tension of the wire. The control and program console also manages the supply, feeding, tension, and speed of the wire. The control and program console of the wiresaw is shown at the left of the photo in Figure 4.6.

4.5.2 The Wire Management Unit

The wire management unit of a wiresaw is illustrated in Figure 4.7. A single cold-drawn steel wire with a typical length of 100 km or longer is supplied at the "feed spool," providing the wire for the entire slicing operation until the ingot is completely sliced through. It is important that the feed spool can supply a long enough wire without re-spooling; otherwise, a visible tool mark shift from one plane to the other is likely to form on the wafer surface. This often results in the entire batch of wafers being discarded, or at least causing undesirable complications in the subsequent lapping and grinding processes. This wire moves through a tension control unit, sometimes called a dancer unit. The tension control unit employs a simple PID controller to ensure that the tension of the feed wire is maintained at a prescribed level before joining the wire web. The tension is typically maintained

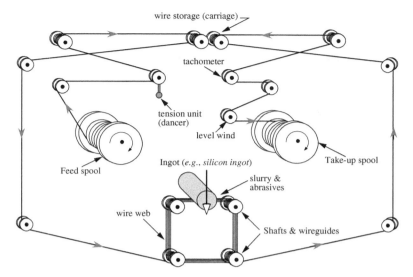

Figure 4.7 Schematic of a wiresaw and illustration of wire management and tension control; the zoom-in view of the wire web and slurry manifold with ingot is illustrated in Figure 4.8

and controlled at a specific value for a given operation, in the range of 20 to 35 N[2] for a steel wire of 150–175 µm in diameter.

The wire web is supported by wire guides, normally in a group of four or three; a typical configuration is shown in Figures 4.6 and 4.7. A more detailed view of the four wire guides, slurry manifold, and the ingot pressing upon the surface of the wire web is illustrated in Figure 4.8. In this figure, the two wire guides on the top are commonly referred to as a master–slave drive pair with a master drive cylinder and a slave cylinder. The master wire guide cylinder drives the wire web to move the parallel wire segments in the longitudinal direction of the wire with a prescribed wire speed. The other two wire guides at the bottom of the figure are followers. Although wire guides are typically arranged in a configuration of four cylinders, they can be arranged in a configuration of two wire guides with a master–slave pair, as illustrated in Figure 4.9(a), or in a configuration of three wire guides as a triangular configuration, as illustrated in Figure 4.9(b), with the top two as the master–slave drive pair and the bottom one as a follower.

Wire guide is made of stainless steel cylindrical drum with surface coated by polyurethane, machined with grooves of constant pitch distance between neighboring grooves. Such grooves constrain the wire segments of the wire web with separation of constant pitch spacing to maintain consistent thickness of wafers during wiresawing operation.

---

2 For a typical cold-drawn steel wire of diameter 175 µm, a tension of 20 N corresponds to approximately 50% of the yield strength. If the wiresaw is operating in a reciprocating mode, the tension of the wire can jump up 10–50% when it reverses the direction, depending on the control algorithm. This means that the wire can reach the verge of rupture if not controlled properly.

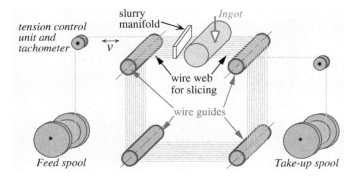

Figure 4.8 A schematic illustration of an ingot on a wire web formed by the four wire guides with a slurry manifold; see also Figure 4.7.

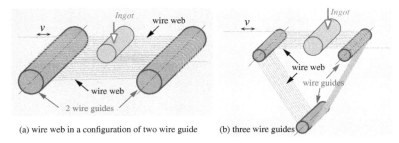

(a) wire web in a configuration of two wire guide   (b) three wire guides

Figure 4.9 Wire web spun with a configuration of (a) two-wire guide cylinders, with master–slave drive control, or (b) three-wire guide cylinders, with the top two as the master–slave drive pair and the bottom one as a follower.

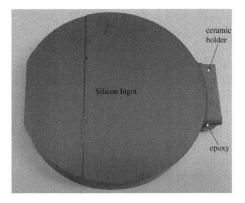

Figure 4.10 An ingot glued to a ceramic fixture base using an epoxy glue. Source: Imin Kao.

As the wire is wound through the wire guides on the grooves sequentially, it creates a net of parallel wire segments known as the "wire web" through which the ingot is fed along with abrasive slurry to produce wafers. The photo in Figure 4.6 shows a silicon ingot mounted on a fixture being fed top-down onto the top horizontal face (not visible) of the wire web for slicing. A photo of an ingot mounted on a ceramic base, glued by epoxy glue, for slicing using a wiresaw is illustrated in Figure 4.10.

Figure 4.11 An illustration of the thickness of a wafer, as related to the wire diameter, kerf loss, and pitch of the wire guides; see Equation (4.1).

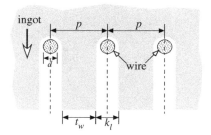

Analysis of kerf loss and wafer thickness based on pitch

The wire guides are coated with a layer of polyurethane or similar polymer materials, and machined with grooves to house and guide each wire segment with equal and prescribed spacing, called the pitch of the wire guide, p. This pitch, representing the spacing between adjacent grooves on the wire guide (or the distance between two adjacent wires on the wire web), determines the thickness of the as-sliced wafers, given by the following equation

$$t_{\mathrm{w}} = p - k_{\mathrm{l}} \tag{4.1}$$

where t_{w} is the thickness of the as-sliced wafer, p is the pitch, and k_{l} is the kerf loss of the slicing process. Figure 4.11 illustrates the parameters of an ingot being fed top-down onto a wire web, showing only three wires. The diameter of the wire is d and $d < k_{\mathrm{l}}$. The total kerf loss, related to the channel through which each wire slices, is the sum of the diameter of the wire d, the clearance of the free abrasives, and the amplitude of vibration of the wire, as expressed in the following equation

$$k_{\mathrm{l}} = d + t_{\mathrm{c}} \tag{4.2}$$

where k_{l} is the kerf loss in a slicing operation using a wiresaw, d is the diameter of the wire, and t_{c} represents the clearance formed around the wire. This can be seen in Figure 4.11 in the channel between the wire and the two walls between the two adjacent wafer surfaces. The clearance t_{c} depends on the average size of the abrasive grits in the slurry and the vibration of the wire between the two walls of the channel.

Since the wire with slurry will also cut the polymer grooves of the wire guides, it is important to inspect periodically and ensure that the grooves remain in working condition without deep channels, which may catch the wire and rupture it. The number of wafers produced in a batch of slicing operation depends on the size of the wire guides and the number of grooves on the wire guides. In some configurations of the slicing operation, multiple ingots can be sliced simultaneously on the same surface of the wire web. Figure 4.12 shows two ingots fed into the wire web surface for slicing. An analytical investigation presented in Wei and Kao (2004) illustrated, through the compliance influence function, that less kerf loss and better surface finish can be obtained by arranging ingots side by side, as in Figure 4.12. Such benefits are achieved due to the increase of the effective stiffness of the wire by moving ingots closer to the boundaries of the wire web in wiresaw slicing process.

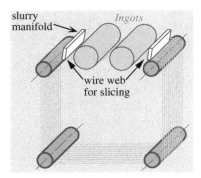

slurry
manifold

Ingots

wire web
for slicing

Figure 4.12 An illustration showing two ingots been fed unto the wire web surface for slicing. Multiple ingots can be fed with appropriate slurry manifold arrangement. Better slicing outcomes can be obtained from such configuration due to the increase of effective stiffness of the system.

Figure 4.7 also illustrates a tachometer unit to measure the speed of the wire for the control of the wire speed through the master drive. The speed of the wire typically ranges from 10 to 15 m s^{-1}.[3]

Once the wire leaves the wire web, it moves along the path and is gathered at the "take-up spool," as shown in Figure 4.7. The take-up spool gathers the used, or spent, wire and levels the wire while winding it on the surface of the take-up spool. The spent wire is of no further use, and is normally recycled.

4.5.3 Uni-directional Versus Bi-directional Wire Motion

The operation of a wiresaw can be broken into two categories, according to the direction of motion of the wire. They are called "uni-directional" and "bi-directional" operations, as will be described in the following. The two different operations affect the total time of slicing, which is an important index of the slicing operation. The total time of slicing obtained based on the length of the wire should be longer than the total time required to slice through the entire crystal ingot without stopping for re-spooling. When a slicing process of an ingot is stopped due to wire breakage or re-spooling of wire, a distinct discontinuity will appear on the surface of all wafers of the ingot with micro-scale steps. This will result in either the wafers being discarded or will require additional effort in the lapping and polishing processes. Therefore, it is important to manage the configuration of the directionality of the wire to ensure that the slicing operation will be completed without re-spooling. The two configurations of wire directionality and the total time of slicing will be presented as follows.

1. **Uni-directional:** As illustrated in the schematic of the wiresaw in Figure 4.7, the wire is fed from the direction of the supply spool to the take-up spool, as shown by the arrows. The wire web spanned by the single wire, winding hundreds of times on the four shafts and wire guides, constitutes the cutting surface of the ingot as it is fed onto the wire web plane.

   The total time of slicing based on the length of the wire is

   $$T_{\text{total}} = \frac{L_s - L_w}{V} \tag{4.3}$$

   where L_s is the total length of the wire on the feed spool in meters, L_w is the wire length between the feed and take-up spools in meters, and V is the speed of

---

3 A speed at 10 m s^{-1} is comparable to the average speed of an Olympic 100 m runner. Thus, the speed of the moving wire of a wiresaw is, on average, typically faster than a competent 100 m runner.

the moving wire in meters per second. The length L_w includes the wire length of the winding and wire web. The length of the wire between the feed and take-up spools, L_w in Equation (4.3), can be neglected, as the length of wire on the wire guides is much shorter than the entire spool of wire; that is,

$$L_w \ll L_s.$$

Thus, Equation (4.3) can be re-written to estimate the total time of slicing with unidirectional wire feeding motion as follows

$$T_{total} = \frac{L_s}{V}. \tag{4.4}$$

In general, the spool of the cold-drawn wire typically used in a wiresaw operation is measured in kilometers, and the speed of the wire in meters per second. Equation (4.4) should be applied by using consistent units, for example the MKS units[4].

Example 4.5.1
A feed spool of wiresaw is supplied with a total of 100 km length of wire. The wire is configured to produce a wire web and moves at a speed of 10 m s^{-1} uni-directionally. Determine the total time it takes to use up the entire spool of wire if the wiresaw operates continuously with the wire moving uni-directionally. Neglect the length of the wire web and the portion of the wire at the take-up spool.

Solution
By neglecting the length of the wire web and the portion of the wire at the take-up spool, the total time it takes to use up the entire spool of wire, according to Equation (4.4), is

$$T_{total} = \frac{100 \text{ km} \times 10^3 \text{m km}^{-1}}{10 \text{ m s}^{-1}} = 10,000 \text{ s} = 2.78 \text{ h.} \tag{4.5}$$

Equation (4.5) indicates that such a spool can continuously supply wire for a slicing operation for about 2.78 h.

Based on the results obtained in Example 4.5.1, the total time of wiresaw operation is about 2.78 h before the wire runs out. If the time required to slice through an ingot is longer than this time duration obtained through Equation (4.5) in Example 4.5.1, the slicing operation will be interrupted when the wire runs out, and will require re-spooling with a new wire spool. This interruption of wiresaw operation will result in a micro-scale step on the wafer surface at the re-spooling interface. This is highly undesirable and can result in scrapping the ingot and unfinished wafers altogether.

2. **Bi-directional:** As suggested in the schematic of the wiresaw in Figure 4.8, the wire can move in both directions with proper wire management and control. The total time of slicing for a given length of wire can be increased when using the bi-directional motion.

In contrast to the setup of the uni-directional feeding of wire, a reciprocating configuration of wire motion can allow the wire to move in both directions in a slicing operation. In a bi-directional wire feeding motion, the wire is fed forward from the

---

4 MKS is a SI unit system using meters for length, kilograms for mass, and seconds for time.

direction of the feed spool to the take-up spool for a certain feed-forward length, L_f, and reverses the direction for a length of L_r in a prescribed unit of reciprocating cycle of motion, where $L_r < L_f$. We define q in the following

$$q = \frac{L_r}{L_f} \times 100\% \qquad (4.6)$$

as the percentage of the reusable wire length in a prescribed reciprocating cycle of wire motion. This ratio q is also called the "B/F ratio". This results in $(100 - q)\%$ of the wire being retired from the slicing operation at any given time, on average. Thus, the same wire spool can be used for $\left(\frac{100}{100-q}\right)$ times longer. If the total length of the wire in the feed spool can supply for a total time of T_{total} in a uni-directional configuration, as calculated in Equation (4.4), then the total time with a bi-direction wire feeding using the above process parameters will be

$$T_{bi} = T_{total} \left(\frac{100}{100 - q}\right). \qquad (4.7)$$

The term in parentheses that multiples T_{total} in Equation (4.7) determines the multiplication factor of the total amount of time that the same spool of wire can be used in slicing operation between the bi-directional and the uni-directional wire management.

Example 4.5.2
As an illustration, we shall use the same parameters and assumption in Example 4.5.1, with a total length of the wire in the feed spool being $L_s = 100$ km. The feed-forward and the reversing lengths for the bi-directional operation are $L_f = 100$ m and $L_r = 80$ m for a prescribed reciprocating cycle with approximately 10 s in forward motion and 8 s in reverse motion, respectively, at a wire speed of 10 m s^{-1}. Determine the total time it takes to use up the entire spool of wire with the bi-directional operation.

Solution
According to Equation (4.6), the percentage of the reusable wire length, or the B/F ratio, is

$$q = \frac{80 \text{ m}}{100 \text{ m}} \times 100 = 80\%$$

The time of slicing using the uni-directional feeding from Equation (4.4) is $T_{total} = 2.78$ h in Example 4.5.1. With the reciprocating wire management, using the prescribed parameters, we obtain the total time of bi-directional slicing operation from Equation (4.7)

$$T_{bi} = 2.78 \times \left(\frac{100}{100 - 80}\right) = 13.9 \text{ h}.$$

Thus, the same spool of wire for a bi-directional process will last for 13.9 h. ∎

By utilizing the bi-directional operation, the same spool of wire will last for 13.9 h, five times longer than that of uni-directional wire operation processes. This will provide enough time to slice through a 200 mm or 300 mm ingot using a wire spool of the same length of wire.

Table 4.1 Comparisons between uni-directional and bi-directional wire motion in wiresaw operation.

Property	Uni-directional motion	Bi-directional motion
As-sliced wafer surface	Better than bi-directional	Good
Time of slicing	Determined by wire length (Equation 4.4)	Longer (Equation 4.7)
Diameter/size of ingots	Small size	Small and large sizes
Tension control	Uniform	Spikes in tension when reversing direction

It takes about 2 to 2.5 h for a typical wiresaw to slice through a polycrystalline silicon ingot with square cross section of 100 mm (about 4 in) length on each side. Thus, a wiresaw with uni-directional wire feeding motion will be able to slice through such ingots without re-spooling. However, a cylindrical single crystalline silicon ingot of a diameter of 200 mm (8 in) will typically require 6–7 h to slice through. A wiresaw with uni-directional wire feeding will not be able to slice through the entire ingot with a spool of 100 km wire. This poses a major drawback of wiresaw operation with a uni-directional wire feeding operation. Thus, a wire management that enables the wire to move in reciprocating motion is necessary in order to lengthen the total time of slicing using a given wire length.

In the early days of wiresaws used in PV wafer slicing, the wire feeding was uni-directional, which produces better as-sliced wafer surfaces. This is suitable for PV wafers with no lapping required after slicing. The bi-directional wire motion becomes necessary as the wafer size increases. Table 4.1 compares the properties between the two categories of wire motions.

4.5.4 The Slicing Compartment

The slicing compartment of the wire web and ingot is illustrated in Figure 4.6, with one side of the wire web circled. The ingot is mounted on a moving fixture support. This compartment is open in the photo for illustration. When the wiresaw is in actual operation, the door of this compartment is closed in order to prevent the abrasive slurry fluid from splashing everywhere. The wire web and guides in Figure 4.7 are further illustrated in a schematic in Figure 4.8, as well as two different configurations in Figure 4.9.

As the ingot is pushed downward onto the surface of the wire web, each wire maintains a bow angle, α, with the cutting surface (the straight line $\overline{AB}$) applying a normal load along the direction of feed (the direction that is perpendicular to $\overline{AB}$) on the contact interface (the curve A-C-B) between the wire and the ingot. Under such a bow, the wire with constant tension will take a position of symmetry corresponding to minimum energy. As such, the cut profile in the wire axial direction can be assumed to be a near circular arc of large radius of curvature.

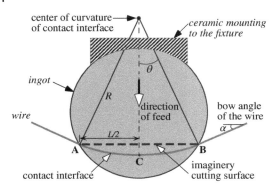

Figure 4.13 Illustration of a wire entering a ingot, forming a bow angle with the imaginary cutting surface to apply a normal load on the contact interface.

Figure 4.13 illustrates the interaction between the ingot and the wire web. In this figure, the ingot is shown to be mounted on a ceramic mounting, glued to the ingot via an epoxy glue which will soften at about 100 °F in order for the wafers to be released from the mounting after the slicing is completed. The wire bow angle, α, varies as the wire web traverses through the ingot, depending on the length of the cutting surface $\overline{AB}$ and the feed rate of the ingot. The bow angle is a result of the feed rate of the ingot and the process parameters. It varies from 5° (0.08727 rad) to 15° (0.2618 rad). This wire bow angle causes a component of the wire tension to become the normal load onto the contact interface. The normal force, N, applied on the ingot by each wire segment of the wire web is

$$N = 2\,P\,\sin\alpha \cong 2\,P\,\alpha \tag{4.8}$$

where P is the tension of the wire, and α is the bow angle in radians. For small angle of α within the typical range, Equation (4.8) can be used when α is in radians. Such normal force interacts with the elasto-hydrodynamic effect to apply pressure on the contact interface, causing the rolling-indenting action of the free abrasive grits in a three-body abrasion to remove material from the ingot surface along the contact interface. The details will be addressed in the modeling of the wiresaw manufacturing process in Chapter 5.

4.5.5 Directions of Ingot Feeding

In the typical design of a wiresaw machine tools, it is easier to move the ingot onto the wire web, instead of moving the entire assembly of wire guides and wire web towards the ingot. This will reduce the amount of tolerance and uncertainty in machining due to the movement of the assembly of wire guides that contain the fast moving wire and the wire web.

As an ingot is fed unto the wire web for slicing, it can be configured to move relative to the wire web in four ways, as follows. The first three configurations of ingot feeding are illustrated in Figure 4.14.

1. Top-down, in which the ingot moves along the direction of gravity.
2. Bottom-up, in which the ingot moves in the opposite direction of gravity.
3. Side-ways, in which the ingot moves in a direction perpendicular to gravity.
4. A combination of the above.

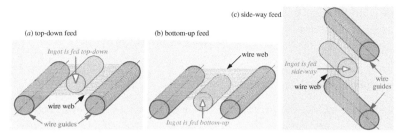

Figure 4.14 Different configurations of feeding an ingot onto the wire web for slicing: (a) top-down ingot feeding, (b) bottom-up ingot feeding, and (c) side-way ingot feeding.

The bottom-up configuration, as illustrated in Figure 4.14(b), was used in the early days of wiresaw machine tools. This might be a reminiscence of the ID saw operation in which the ID blade often traversed down toward the ingot for slicing – equivalent to the ingot moving relative to the ID blade in a direction against the direction of gravity. However, nearly all wiresaws today employ the top-down configuration of ingot feeding, as illustrated in in Figure 4.14(a). There are a few reasons for this preferred configuration of ingot feeding. First of all, it is better to have the moving mechanism of ingot feeding outside of, and separate from, the assembly of the wire guides and wire web. A top-down ingot feeding allows this to happen easily, as opposed to the bottom-up mechanism in which the ingot and fixtures must fit within the enclosure space formed by the wire guides. A bottom-up configuration also requires enough space within the enclosure, making it impossible to have wire guides arranged in ways that save space and operation costs, such as those in Figure 4.9.

More importantly, the sliced wafers of the top-down configuration hang like pendulums during the slicing process, with gravity helping to stabilize the wafers against the vibration caused by the machining process. Such stability reduces the likelihood of premature breakage of wafers during the slicing process. On the other hand, the sliced wafers of the bottom-up configuration are similar to the inverted pendulums, which is inherently unstable. With induced vibration from the manufacturing process, the wafers of the bottom-up configuration are more prone to premature breakage during the slicing process.

In addition, the fixture and glue that affix the ingot must be able to secure the ingot and wafers throughout the slicing process, especially in a top-down configuration because the force of gravity and slicing tend to pull the ingot away from its base.

The third configuration arranges ingot feeding side ways, as illustrated in Figure 4.14(c). The sliced wafers in this configuration are akin to cantilever beams. They are more susceptible to lateral vibration, perpendicular to the direction of the wafer surface. This is due to the smaller modulus of rigidity in beam bending in the lateral direction compared with that in the direction along the surface. The stability of the sliced wafers, subject to vibration, is between that of the top-down and bottom-up configurations.

The fourth configuration is an appropriate combination of the top-down and side-ways configurations, or the bottom-up and side-way configurations. Such configurations may be applicable in specially customized situations.

4.5.6 Consumables and Other Operations

The two main consumables in a wiresaw operation are the wire and abrasive slurry. The more expensive consumable of the two is the slurry, consisting of the abrasive grits and the carrier, as described in Section 3.8.

The wire used in the wiresawing process is made by the wire drawing process called the "cold-drawn" (CD) process – a very standard industrial process for metal forming [Groover (2012); Kalpakjian and Schmid (2013)]. The cold-drawn process improves the yield strength of the wire due to the nature of the manufacturing process. This increased yield strength is a very desirable property in slicing using the cold-drawn wires. The wire drawing manufacturing process produces wire of the same diameter with consistent quality and nearly no defects. In a typical slicing operation with well controlled process parameters and slurry supply, the wire hardly breaks. These types of wire have been used in other applications, such as in the manufacture of radial tires; thus, the wire is very inexpensive. In comparison with the abrasive slurry, the cost of the wire as a consumable in a wiresaw operation is almost negligible.

The main cost of slurry is the abrasive grits. Silicon carbide abrasive grits are usually used in slicing silicon wafers using wiresaws. Each batch of new SiC abrasives and carrier fluid is typically used five times before being recycled or discarded. Recycling of abrasive grits in a wiresaw operation has been shown to save cost and produce a better surface finish (cf. Section 3.8).

Crystal ingots are glued onto a fixture mounted on the feeding mechanism to feed the ingot onto the wire web. The ingot is glued to the fixture using an epoxy, which softens at an elevated temperature in a water bath after slicing is completed such that the sliced wafers may be easily detached from the fixture without breakage.

4.6 Comparison Between the ID Saw and Wiresaw

The comparison of the process parameters and outcomes between the wiresaw and ID saw is presented in this section. First of all, the schematic drawings of the two machine tools are compared side by side in Figure 4.15. The wiresaw consists of one wire forming a wire web, traversing through a silicon ingot, as shown in Figure 4.15(a). A single wire is wound carefully on the wire guides with grooves of constant pitch to form a wire web. A pair of master-slave drives rotate the wire guides, causing the entire wire web to move at high speed while carrying the abrasive slurry to remove the material from the contact interface with the ingot. The ingot is fed in the direction perpendicular to the wire web, as shown by the arrow in Figure 4.15(a). The wire maintains a constant tension during the cutting process. The schematic of the ID saw, as illustrated in Figure 4.15(b), shows an ID

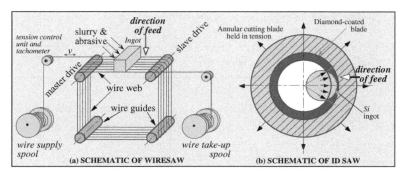

Figure 4.15 Schematic diagrams and comparison of (a) the modern wiresaw, and (b) an ID saw.

blade moving towards the ingot mounted on a stationary fixture to slice one wafer at a time. The direction of the feeding motion of the ID blade is shown by the arrow in Figure 4.15(b).

Modern wiresaw manufacturing is based on the so-called "free abrasive machining" (FAM) process which removes materials via third-party free abrasives in slurry, whereas the machining of the conventional ID saw is based on the "ploughing machining" process generated with bonded abrasive grits, characterized by the removal of material through forceful agents attached to the tool.

Depending upon the process control, different levels of parallel wire marks may be visible on the wafer surface after they are sliced. A wiresaw is capable of slicing wafers of large diameter as long as the distance between the wire guides is larger than the diameter of the ingot. With a wire web consisting of 200–400 strands of wire segments wound over the wire guides, the same total number of 200–400 wafers are produced, corresponding to the total number of strands of wire segments in the wire web. All wafers are simultaneously sliced through once the ingot has finished traversing through the wire web. This takes between six and eight hours for a 200 mm silicon ingot. Typical kerf loss is dependent upon the diameter of the wire, the average size of abrasives, and the vibration amplitude of the wire. The thickness of the wafer is controlled by the pitch of the grooves on the wire guides.

The kerf loss of a wafer sliced by the ID saw is the sum of the thickness of the ID blade, the thickness of the coated diamond surface, and the amplitude of vibration of the blade as it slices through the ingot, as expressed in Equation (4.9)

$$k_1 = t_{ID} + t_{ab} + t_{vib} \tag{4.9}$$

where k_1 is the kerf loss in a slicing operation using an ID saw, t_{ID} is the thickness of the ID blade, t_{ab} is the average size of the diamond abrasive grains (also see Section 3.5) and the bonding above the surface of the ID blade, and t_{vib} is the amplitude of vibration of the ID blade during machining. Due to the bonded abrasive machining process, the surface of a wafer sliced by ID saws will have more surface fractures and damages, shown in Figure 4.4, as compared to that of a wafer sliced by wiresaw. Comparisons of various properties between the wiresaw and ID saw are offered in Table 4.2.

Table 4.2 Comparison of various properties between the modern slurry wiresaw and the ID saw.

Property	Wiresaw	ID saw
Cutting method	FAM/lapping	BAM/ploughing
Typical cut surface features	Wire marks	Chipping and fracture
Surface fractures and damage	Less	Significant
Slice single- or poly-crystalline ingots	Yes	Yes
Depth of subsurface damage (SSD)	Uniform 5–15 μm	Variable 20–35 μm
Productivity	110–220 cm^2 h^{-1}	10–30 cm^2 h^{-1}
Wafers per run	200–400 wafers	One wafer
Kerf loss	180–210 μm	300–500 μm
Minimum thickness of wafer	200 μm (typ.)	350 μm
Yield of 700 μm thick wafers	220 per 20 cm	180 per 20 cm
Maximum ingot diameter	300 mm and higher	Up to 200 mm
Number of wafers after each cut	Hundreds	One
Moving part	Ingot	ID saw blade

Table 4.2 compares the wiresaw with the ID saw, pertaining to the various properties of the wafers sliced by each machine tool and the process parameters. The as-sliced wafers of wiresaw have regular tool marks caused by the wire and random pits due to the FAM process, as compared to the as-sliced wafers of an ID saw which have visible fractures and damage of brittle chipping on the surface. Wafers sliced by ID saws have larger depth of SSD due to the BAM ploughing process, requiring more material removal from the wafer surface in the subsequent lapping and polishing processes.

Both wiresaws and ID saws can slice single crystalline or polycrystalline silicon ingots with round or rectangular cross-section. Both saws can also slice through highly anisotropic crystals (such as lithium niobate) effectively. Note that silicon is also anisotropic.

Several process parameters are compared in the table. It is clear that the wiresaw has much higher throughput and yield with less kerf loss than the ID saw. The productivity, measured in the total output of wafer surface area per hour, of a wiresaw is higher than that of the ID saw. Wiresaws also produce from ingots more wafers per each run. With well-controlled slicing processes, wiresaws can slice very thin wafers at 150–200 μm, useful for PV applications to reduce materials cost.

In the following, a few examples are used to discuss the throughput of wafer slicing using a wiresaw versus an ID saw.

Example 4.6.1

A silicon ingot of $L = 300$ mm in length is to be sliced by a wiresaw. The average kerf loss is found to be 185 μm using this wiresaw machine tool. The required thickness of the as-sliced wafer is 750 µm. Determine the the number of strands of wires needed for the wire web. How many wafers can be produced after the wire web slices through the silicon ingot?

Solution
Refer to Figure 4.11 and Equation (4.1); the required pitch of the wire web is

$$p = t_w + k_1 = 750 + 185 = 935 \ \mu\text{m}$$

where $t_w = 750$ µm is the desired thickness of the wafer, and $k_1 = 185$ µm is the given kerf loss. With the pitch of the wire web, the estimated number of wafer slices is

$$N = \frac{L}{p} = \frac{300 \text{ mm}}{935 \ \mu\text{m}} = 320.8.$$

A total of about 320 wafers can be produced, assuming no loss due to the removal of silicon wafers from the epoxy glue on the ingot holder after slicing, and no wafers break due to handling.

In addition, due to the possible uneven geometry at the ends of the ingot, the total number of strands of wire of the wire web should be about 326 to cover the entire range of the 300 mm of silicon ingot. The total number of wafers produced is about 320. The actual number of slices of wafer may be a little fewer than 320 if the geometric irregularity at both ends of the ingot causes loss of a couple of wafers at the edge.

Example 4.6.2
The wiresaw in Example 4.6.1 is replace by an ID saw, which has an average kerf loss of 400 µm. The required thickness of the as-sliced wafer is 750 µm. How many wafers can be produced after the ID saw slicing operation is completed, with the same silicon ingot of length $L = 300$ mm?

Solution
Refer to equation (4.1); the pitch of slicing for the ID saw is

$$p = t_w + k_1 = 750 + 400 = 1150 \ \mu\text{m}$$

where $t_w = 750$ µm is the desired thickness of the wafer, and $k_1 = 400$ µm is the kerf loss of this ID saw. With the pitch of wafer slicing, the estimated number of wafer slices is

$$N = \frac{L}{p} = \frac{300 \text{ mm}}{1.15 \text{ mm}} = 260.9.$$

Thus, a total of about 260 wafers can be produced by this ID saw with the same ingot of length $L = 300$ mm, assuming no loss during the machining and handling processes.

A quick comparison between Examples 4.6.1 and 4.6.2 reveals that the wiresaw can produce about 60 more wafer slices, a 23% increase in the throughput of wafer numbers produced. This difference is primarily due to the larger kerf loss of the ID saw.

When slicing polycrystalline silicon ingots to produce photovoltaic (PV) wafers, the desired thickness of wafers is typically much smaller than the thickness of a single crystalline silicon wafer for IC device fabrication. PV wafers are typically processed to make solar panels without surface processing, such as lapping; they are usually processed to make solar panels as-sliced. Examples in the following are used to compare the throughput of PV wafers using both an ID saw and a wiresaw.

Example 4.6.3
A polycrystalline silicon ingot of $L = 300$ mm in length, with a square cross section, is to be sliced to produce PV wafers. The average kerf loss is 185 µm for the wiresaw and 400 µm for the ID saw. The required thickness of the as-sliced wafer is 350 µm. How many PV wafers can be produced by the wiresaw and ID saw on the same ingot?

Solution
Refer to equation (4.1), the required pitches of for the wiresaw and ID saw are

$$p_{\mathrm{w}} = t_{\mathrm{w}} + k_l = 350 + 185 = 535 \ \mu\mathrm{m}$$
$$p_{\mathrm{id}} = t_{\mathrm{w}} + k_l = 350 + 400 = 750 \ \mu\mathrm{m}$$

where $t_{\mathrm{w}} = 350$ µm is the desired thickness of the PV wafers, with the corresponding kerf losses. With the pitches p_{w} and p_{id}, the estimated numbers of wafer slices are

$$N_{\mathrm{w}} = \frac{L}{p_{\mathrm{w}}} = \frac{300 \ \mathrm{mm}}{535 \ \mu\mathrm{m}} = 560.7$$
$$N_{\mathrm{id}} = \frac{L}{p_{\mathrm{id}}} = \frac{300 \ \mathrm{mm}}{750 \ \mu\mathrm{m}} = 400.$$

A total of about 560 PV wafers can be produced by the wiresaw, while a total of 400 PV wafers can be produced by the ID saw, assuming no loss during the machining and handling processes.

Example 4.6.3 illustrates that an additional 160 PV wafers can be produced by the wiresaw, equivalent to an increase of throughput by 40%. This difference is due to the smaller thickness of PV wafers, in addition to the larger kerf loss of the ID saw.

4.7 Research Issues in Wiresaw Manufacturing Processes

Many research works have been conducted with results reported by both academia and industry for understanding, modeling, and improving the process of wiresaw manufacturing. These results represent significant advancement in the modeling and understanding of the wiresawing process since the 1990s. References listed for further reading include the following: the rolling-indenting mechanism of machining in Bhagavat and Kao (1999); Bhagavat and Kao (2005,2006b); Kao (2004); Li et al. (1998); Moller (2004); Yang and Kao (1999); Yang and Kao (2001), investigation of the physics of hydrodynamic interaction in Bhagavat et al. (2000); Bierwisch et al. (2011); Moller (2006); Zhu and Kao (2005a,b, n.d.); Zhu et al. (2000), warp due to thermal expansion in Abe et al. (2009); Bhagavat et al. (2003); Bhagavat and Kao (2008); Johnsen et al. (2012); Yamada et al. (2002), vibration of an axially moving wire in Chung and Kao (2008a,b,2011a,b); Kao et al. (1998d); Mizuno et al. (2006); Wei and Kao (2000a,b); Wei and Kao (2004); Zhu and Kao (2005b); Zhu et al. (2000), crystal anisotropy in Bhagavat and Kao (2006a), modeling of the interaction among abrasives in Bierwisch et al. (2011); Liedke and Kuna (2013); Nassauer and Kuna (2013); Nassauer et al. (2014); Nassauer et al. (2013), macroscopic wiresawing model in Liedke and Kuna (2011), recycle and debris of slurry in Bidiville et al. (2011, 2010); Costantini et al. (2000); Fallavollita (2011); Kozasa and Gotou (2009); Lan et al. (2011); Neece (2006); Watson and Glaser (1966), loss of wire tension in Meiner et al. (2014), slurry composition in Bidiville et al. (2010); Chandra et al. (1998); Skomedal et al. (2011), and others Bhagavat et al. (1998); Bhagavat and Kao (2004); Bhagavat et al. (2005); Bhagavat et al. (2010); Gouldstone and Kao (2006); Kao (2001); Kao (2004); Kao (2005a,b); Kao and Bhagavat (2007); Kao et al. (1998a,b,c); Kao et al. (2010); Kao et al. (1997); Li et al. (1997); Li et al. (1998); Nasch and Sumi (2009); Sahoo et al. (1996); Sahoo et al. (1998); Wei and Kao (2004); Yang and Kao (2001); Yang et al. (1999). A good review of wiresawing technology can be found in Wu (2016).

4.8 Summary

In this chapter, the wiresaw is introduced as a machine tool for wafer slicing. The historical perspectives of saws using wires are presented, including wire saws used for quarrying, slabbing of granite blocks, and concrete cutting, as well as the carpenters saw for woodworking. The rise of the PV industry in the early 1990s had a profound impact on the emergence of the modern slurry wiresaw for wafer slicing, replacing ID saws as the primary tool for wafer slicing. Since the late 1990s, wiresaws have been widely adopted for the slicing of semiconductor silicon wafers. Three categories of saws for wafer slicing were introduced in this chapter – the ID saw, the wiresaw, and saws with wires coated with or impregnated by diamond abrasive grits. The various components and operations of the wiresaw machine tool were presented in detail, including the wire guides, control, wire management, wire web, uni- and bi-directional wire feeding, ingot feeding, slicing one ingot or multiple ingots on one wire web, slurry manifold, and consumables for wiresaw operation, among others. Comparison between an ID saw and a wiresaw was offered with discussions pertaining to the various properties of the wafers sliced by each machine tool and the process parameters. Finally, SEM photos of wafers sliced by a wiresaw and an ID saw were presented and contrasted. Research investigations on various aspects of wiresaw manufacturing process are listed for further reading.

References

Abe Y, Ishikawa KI and Suwabe H 2009 Effects of thermal deformation of multi-wire saw's wire guides and ingot on slicing accuracy. *Key Engineering Materials* **389–390**, 442–447.

Bhagavat M and Kao I 1999 Computational model for free abrasive machining of brittle silicon using a wiresaw *the Proceedings of IMECE'99: DE-Vol 104, Electronics Manufacturing Issues*, pp. 21–30. ASME Press.

Bhagavat M, Prasad V and Kao I 2000 Elasto-hydrodynamic interaction in the free abrasive wafer slicing using a wiresaw: Modeling and finite element analysis. *Journal of Tribology* **122**(2), 394–404.

Bhagavat M, Wei S, Yang F, Kao I and Prasad V 1998 Modeling and control of wiresaw slicing *the 12th International Conference on Crystal Growth*, Jerusalem, Israel.

Bhagavat M, Witte D, Kimbel S, Sager D and Peyton J 2003 Method and apparatus for slicing semiconductor wafers. US Patent App. 10/377,271.

Bhagavat S and Kao I 2004 Nanoindentation studies on a non-centrosymmetric crystal: Lithium niobate *Invited talk at the High Pressure Phase Transformation Workshop, NSF Focused Research Group (FRG)*, Raleigh, North Carolina State University.

Bhagavat S and Kao I 2005 Nanoindentation of lithium niobate: Hardness anisotropy and pop-in phenomenon. *Materials Science and Engineering A: Structural Materials: Properties, Microstructure and Processing* **393**, 327–331.

Bhagavat S and Kao I 2006a Theoretical analysis on the effects of crystal anisotropy on wiresawing process and application to wafer slicing. *International Journal of Machine Tools and Manufacture* **46**, 531–541.

Bhagavat S and Kao I 2006b Ultra-low load multiple indentation response of materials: In purview of wiresaw slicing and other free abrasive machining (FAM) processes. *International Journal of Machine Tools and Manufacture* **46**(5), 531–541.

Bhagavat S and Kao I 2008 A finite element analysis of temperature variation in silicon wafers during wiresaw slicing. *International Journal of Machine Tools and Manufacture* **48**, 95–106.

Bhagavat S, Liberato J and Kao I 2005 Effects of mixed abrasive slurries on free abrasive machining processes *Proceedings of the 2005 ASPE Conference* ASPE.

Bhagavat S, Liberato J, Chung C and Kao I 2010 Effects of mixed abrasive grits in slurries on free abrasive machining (FAM) processes. **50**, 843–847.

Bidiville A, Neulist I, Wasmer K and Ballif C 2011 Effect of debris on the silicon wafering for solar cells. *Solar Energy Materials and Solar Cells* **95**(8), 2490–2496.

Bidiville A, Wasmer K, Michler J, Nasch P, der Meer MV and Ballif C 2010 Mechanisms of wafer sawing and impact on wafer properties. *Progress in Photovoltaics: Research and Applications* **18**(8), 563–572.

Bierwisch C, Kubler R, Kleer G and Moseler M 2011 Modelling of contact regimes in wire sawing with dissipative particle dynamics. *Philosophical Transactions of the Royal Society A: Mathematical, Physical and Engineering Sciences* **369**(1945), 2422–2430.

Chandra M, Leyvraz P, Talbott JA, Gupta K, Kao I and Prasad V 1998 Challenges in slicing large diameter silicon wafers using slurry wiresaw *Proceedings of the Manufacturing Engineering Division, IMECE'98*, pp. 807–811. ASME Press.

Chonan S and Hayase T 1987 Stress analysis of a spinning annular disk to a stationary distributed, in-plane edge load. *J. of Vibration Acoustics Stress and Reliability in Design* **107**, 277–282.

Chonan S, Jiang ZW and Yuki Y 1993a Stress analysis of a silicon-wafer slicer cutting the crystal ingot. *Journal of Mechanical Design* **115**, 711–717.

Chonan S, Jiang ZW and Yuki Y 1993b Vibration and deflection of a silicon-wafer slicer cutting the crystal ingot. *Journal of Vibration and Acoustics* **115**, 529–534.

Chung C and Kao I 2008a Comparison of free abrasive machining processes in wafer manufacturing *Proceedings of International Manufacturing Science and Engineering Conference (MSEC 2008)* ASME Paper number MSEC2008-72253 ASME.

Chung C and Kao I 2008b Damped vibration response at different speeds of wire in slurry wiresaw manufacturing operations *Proceedings of International Manufacturing Science and Engineering Conference (MSEC 2008)* ASME Paper number MSEC2008-72213 ASME.

Chung C and Kao I 2011a Green's function and forced vibration response of damped axially moving wire. *Journal of Vibration and Control* **18**(12), 1798–1808.

Chung C and Kao I 2011b Modeling of axially moving wire with damping: Eigenfunctions, orthogonality and applications in slurry wiresaws. *Journal of Sound and Vibration* **300**, 2947–2963.

Clark W, Shih A, Hardin C, Lemaster R and McSpadden S 2003a Fixed abrasive diamond wire machining- part i: Process monitoring and wire tension force. *International Journal of Machine Tools and Manufacture* **43**(5), 523–532.

Clark W, Shih A, Lemaster R and McSpadden S 2003b Fixed abrasive diamond wire machining- part ii: Experiments design and results. *International Journal of Machine Tools and Manufacture* **43**(5), 533–542.

Contardi GL 1993 Wire saw beads. Economic production. *Industrial Diamond Review*.

Costantini M, Talbott J, Chandra M, Prasad V, Caster A, Gupta K and Leyvraz P 2000 Method and apparatus for improved wire saw slurry. US Patent 6,113,473.

Fallavollita JA 2011 Methods and apparatus for recovery of silicon and silicon carbide from spent wafer-sawing slurry. US Patent App. 12/865,989.

Gouldstone A and Kao I 2006 Wafer slicing using slurry wiresaw and relevance of nanoindentation in its analysis *High Pressure Phase Transformation (HPPT) Workshop*, Kalamazoo, Michigan.

Groover MP 2012 *Fundamentals of Modern Manufacturing: Materials, Processes, and Systems* 5th edn. Wiley.

Hayes D 1990 Demolition. the modern method. *Industrial Diamond Review*.

Herbert S 1989 UK's biggest wire saw contract. *Industrial Diamond Review*.

Johnsen L, Olsen JE, Bergstrm T and Gastinger K 2012 Heat transfer during multiwire sawing of silicon wafers. *Journal of Thermal Science and Engineering Applications*.

Kalpakjian S and Schmid S 2013 *Manufacturing Engineering & Technology* 7th edn. Prentice Hall.

Kao I 2001 Research towards the next-generation re-configurable wiresaw for wafer slicing and on-line real-time metrology *in the CD-ROM Proceedings of the 2001 NSF Design, Service and Manufacturing Grantees and Research Conference*, Tempa, Florida.

Kao I 2004 Technology and research of slurry wiresaw manufacturing systems in wafer slicing with free abrasive machining. *the International Journal of Advanced Manufacturing Systems* **7**(2), 7–20.

Kao I 2005a Experiments of nanoindentation on non-centrosymmetric crystal and study of their implication in wafer manufacturing processes In *Invited talk at the US-Africa Workshop on Materials and Mechanics* (ed. sponsored workshop N), Cape Town, South Africa.

Kao I 2005b Experiments of nanoindentation on non-centrosymmetric crystal and study of their implication in wafer manufacturing processes In *the US-Africa Workshop on Materials and Mechanics* (ed. sponsored workshop N), Cape Town, South Africa.

Kao I and Bhagavat S 2007 *Single-crystalline silicon wafer production using wire saw for wafer slicing* Transworld Research Network Editors J. Yan and J. Patten, Kerala, India chapter 7, pp. 243–270.

Kao I, Bhagavat M and Prasad V 1998a Integrated modeling of wiresaw in wafer slicing *NSF Design and Manufacturing Grantees Conference*, pp. 425–426, Monterey, Mexico.

Kao I, Chung C and Rodriguez R 2010 *Modern Wafer Manufacturing and Slicing of Crystalline Ingots to Wafers Using Modern Wiresaw* Springer Handbook of Crystal Growth; *ed.* G. Dhanaraj M. Dudley, K. Byrappa, and V. Prasad Springer-Verlag chapter 52, pp. 1719–1736.

Kao I, Prasad V, Chandra M, Costantini M and Talbott J 1998b Wafer slicing technology for semiconductor and photovoltaic applications *the 12th International Conference on Crystal Growth*, Jerusalem, Israel.

Kao I, Prasad V, Chiang FP, Bhagavat M, Wei S, Chandra M, Costantini M, Leyvraz P, Talbott J and Gupta K 1998c *Modeling and experiments on wiresaw for large silicon wafer manufacturing the 8th Int. Symp. on Silicon Mat. Sci. and Tech.*, p. 320, San Diego.

Kao I, Prasad V, Li J, Bhagavat M, Wei S, Talbott J and Gupta K 1997 Modern wiresaw technology for large crystals *Proceedings of ACCGE/east-97*, Atlantic City, NJ.

Kao I, Wei S and Chiang FP 1998d Vibration of wiresaw manufacturing processes and wafer surface measurement *NSF Design and Manufacturing Grantees Conference*, pp. 427–428, Monterey, Mexico.

Kojima M, Tomizawa A and Takase J 1990 Development of new wafer slicing equipment (unidirectional multi wire-saw). *Sumitomo Metals* **42**(4), 218–224.

Kozasa K and Gotou I 2009 Method of recycling abrasive slurry. US Patent App. 12/192,351.

Lan C, Lin Y, Wang T and Tai Y 2011 Recovery method of silicon slurry. US Patent 8,034,313.

le Scanff A 1988 New wire saw machine. *Industrial Diamond Review.*

Li J, Kao I and Prasad V 1997 *Modeling stresses of contacts in wiresaw slicing of polycrystalline and crystalline ingots: Application to silicon wafer production Proceedings of ASME IMECE '97*, pp. 439–446. ASME Press, Dallas, Texas.

Li J, Kao I and Prasad V 1998 Modeling stresses of contacts in wiresaw slicing of polycrystalline and crystalline ingots: Application to silicon wafer production. *Journal of Electronic Packaging* **120**(2), 123–128.

Liedke T and Kuna M 2011 A macroscopic mechanical model of the wire sawing process. *International Journal of Machine Tools and Manufacture* **51**(9), 711–720.

Liedke T and Kuna M 2013 Discrete element simulation of micromechanical removal processes during wire sawing. *Wear* **304**(1-2), 77–82.

Meiner D, Schoenfelder S, Hurka B, Zeh J, Sunder K, Koepge R, Wagner T, Grun A, Hagel HJ, Moeller HJ, Schwabe H and Anspach O 2014 Loss of wire tension in the wire web during the slurry based multi wire sawing process. *Solar Energy Materials and Solar Cells* **120**(PART A), 346–355.

Mitchell KW, Richard R, Jester TL and McGraw M 1994 Reforming of CZ Si photovoltaics *IEEE Proc. of 24th Photovoltaics Specialists Conference*, pp. 1266–1269.

Mizuno M, Iyama T, Kikuchi S and Zhang B 2006 Development of a device for measuring the transverse motion of a saw-wire. *Journal of Manufacturing Science and Engineering, Transactions of the ASME* **128**, 826–834.

Moller HJ 2004 Basic mechanisms and models of multi-wire sawing. *Advanced Engineering Materials* **6**(7), 501–513.

Moller HJ 2006 Wafering of silicon crystals. *Phys. Stat. Sol.* (a) **203**(4), 659–669.

Nasch PM and Sumi R 2009 Meeting current and future wafering challenges. *Solid State Technology* **52**(3), 32+34–36.

Nassauer B and Kuna M 2013 Contact forces of polyhedral particles in discrete element method. *Granular Matter* **15**(3), 349–355.

Nassauer B, Hess A and Kuna M 2014 Numerical and experimental investigations of micromechanical processes during wire sawing. *International Journal of Solids and Structures* **51**(14), 2656–2665.

Nassauer B, Liedke T and Kuna M 2013 Polyhedral particles for the discrete element method: Geometry representation, contact detection and particle generation. *Granular Matter* **15**(1), 85–93.

Nature 1901 The use of the wire saw for quarrying. *Nature* **65**(1674), 84–84.

Neece T 2006 Review on sic-recycling in wafer sawing operations. *InterCeram: International Ceramic Review* **55**(6), 430–432.

Sahoo R, Prasad V, Kao I, Talbott J and Gupta K 1996 Towards an integrated approach for analysis and design of wafer slicing by a wire saw In *1996 ASME IMECE, Manufacturing Science and Engineering* (ed. Subramania K), pp. 131–140.

Sahoo R, Prasad V, Kao I, Talbott J and Gupta K 1998 Towards an integrated approach for analysis and design of wafer slicing by a wire saw. *Journal of Electronic Packaging* **120**(1), 35–40.

Skomedal G, Ovrelid E, Armada S and Espallargas N 2011 Effect of slurry parameters on material removal rate in multi-wire sawing of silicon wafers: A tribological approach.*Proceedings of the Institution of Mechanical Engineers, Part J: Journal of Engineering Tribology* **225**(10), 1023–1035.

Statista 2014 Renewable energy: global solar PV market size 2013. URL http://www.statista.com/statistics/232859/global-solar-pv-market-size/.

Watson WI and Glaser PW 1966 Silicon carbide recovery. US Patent 3,259,243.

Wei S and Kao I 2000a Free vibration analysis for thin wire of modern wiresaw between sliced wafers in wafer manufacturing processes *the Proc. of IMECE'00: EEP-Vol 28 packaging of electronic and photonic devices*, pp. 213–219. ASME Press, Orlando, Florida.

Wei S and Kao I 2000b Vibration analysis of wire and frequency response in the modern wiresaw manufacturing process. *Journal of Sound and Vibration* **231**(5), 1383–1395.

Wei S and Kao I 2004 Stiffness analysis in wiresaw manufacturing systems for applications in wafer slicing. *the Int. Journal of Advanced Manufacturing Systems* **7**(2), 57–64.

Wu H 2016 Wire sawing technology: A state-of-the-art review. *Precision Engineering* **43**, 1–9.

Yamada T, Fukunaga M, Ichikawa T, Furno K, Makino K and Yokoyama A 2002 Prediction of warping in silicon wafer slicing with wire saw. *Theoretical and Applied Mechanics* **51**, 251–258.

Yang F and Kao I 1999 Interior stress for axisymmetric abrasive indentation in the free abrasive machining process: Slicing silicon wafers with modern wiresaw. *Journal of Electronic Packaging* **121**(3), 191–195.

Yang F and Kao I 2001 Free abrasive machining in slicing brittle materials with wiresaw. *Journal of Electronic Packaging* **123**, 254–259.

Yang F, Li JCM and Kao I 1999 Interaction between ingot and wire in wiresaw process In *the Proceedings of IMECE'99: DE-Vol 104, Electronics Manufacturing Issues* (ed. Sahay C, Sammakia B, Kao I and Baldwin D), pp. 3–8. ASME Press, Three Park Ave., New York, NY 10016.

Zhu L and Kao I 2005a Computational model for the steady-state elasto-hydrodynamic interaction in wafer slicing process using wiresaw. *the Int. Journal of Manufacturing Technology and Management* **7**(5/6), 407–429.

Zhu L and Kao I 2005b Galerkin-based modal analysis on the vibration of wire-slurry system in wafer slicing using wiresaw. *Journal of Sound and Vibration* **283**(3-5), 589–620.

Zhu L and Kao I n.d. Equilibrium elastohydrodynamic interaction analysis in wafer slicing process using wiresaw *Proc. of IMECE'01: EEP*, vol. 1.

Zhu L, Bhagavat M and Kao I 2000 Analysis of the interaction between thin-film fluid hydrodynamics and wire vibration in wafer manufacturing using wiresaw *the Proc. of IMECE'00: EEP-Vol 28 packaging of electronic and photonic devices*, pp. 233–241. ASME Press, Orlando, Florida.

5

Modeling of the Wiresaw Manufacturing Process and Material Characteristics

This chapter continues the presentation from Chapter 4 to address the fundamentals of modern slurry wiresaw machining and engineering modeling. The topics include the following: an introduction of fundamentals of engineering modeling, the rolling-indenting engineering model, analysis of vibration for the moving wire in the wiresaw manufacturing process, modeling of practical damping factors, elasto-hydrodynamic interaction between the high-speed elastic wire, slurry, and the ingot contact surface, thermal management, slurry and wire management, and material properties and their effects on wafering.

5.1 Introduction

Although the manufacturing process of a modern slurry wiresaw is subject to more stringent requirements to produce wafers with low total thickness variation (TTV), low warpage, and low residual stresses, the operation of industrial wiresaws today rely on ad hoc tuning of various process parameters to produce wafers in satisfactory conditions. Understanding the fundamental manufacturing process will enable and facilitate further development and advancement in wiresaw technology with more systematic results. This is the main motivation for the presentation of the recent research results in the "rolling-indenting" engineering modeling of wiresaw manufacturing process, as well as other relevant analyses [Kao et al. (1998a, 1997a,b); Li et al. (1997, 1998); Sahoo et al. (1998); Zhu and Kao (2005b); Kao et al. (1998b); Wu et al. (1997)].

In the following sections, the theoretical modeling and experimental studies of the modern wiresaw manufacturing systems are presented and discussed. The mechanism of brittle machining on the removal of materials using a wiresaw was discussed in Chapter 3. In this section, the fundamental engineering model of "rolling-indenting" of the wiresawing process is described. After that, the analyses of vibration and hydrodynamics of a moving wire with viscous slurry are presented. Vibration analysis of a moving wire is a new research problem that has received attention since the early 1990s. The vibration of the wire segments in the wire web of a wiresaw can have direct impact on the kerf loss and the quality of wafer surface. Furthermore, because of the high speed and low mass density per unit length

Wafer Manufacturing: Shaping of Single Crystal Silicon Wafers,
First Edition. Imin Kao and Chunhui Chung.
© 2021 John Wiley & Sons Ltd. Published 2021 by John Wiley & Sons Ltd.

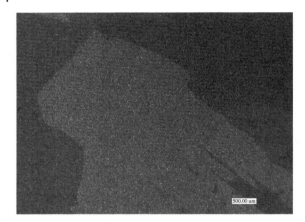

Figure 5.1 Horizontal and parallel tool marks made by the wire of a wiresaw on the surface of a polycrystalline silicon wafer (at 50X magnification). Source: Imin Kao.

of the wire, the hydrodynamic effect takes place and results in a hydrodynamic film gap between the workpiece and the wire, which is typically larger than the size of abrasive grits [Bhagavat et al. (2000); Zhu and Kao (2005a)]. Therefore, the rolling-indenting machining is expected to dominate the wiresaw slicing.

The surface of wafers produced with a well-controlled slicing process should be scratch-free. As an illustration, Figure 5.1 shows the tool marks of an as-sliced poly-crystalline silicon wafer surface. The horizontal and evenly spaced tool marks are the result of a wire moving on the ingot to slice wafers. The grain boundary of the poly-crystalline silicon is visible on the surface, under a 50X magnification. It is noted that the parallel tool marks are produced during the operation of wiresawing in either uni-directional or bi-directional wire management (cf. Section 4.5.2). The tool marks on wafers sliced by a well controlled wiresaw with uni-directional feeding can be less severe than those sliced by bi-directional reciprocating feeding.

Another photo with 500X magnification in Figure 5.2 shows the details of random pits on the surface of an as-sliced wafer, with a scratch-free wafer surface. Such types of surfaces are typical of a well-controlled and managed slicing process using a wire-saw. This is in contrast to surfaces of the wafer in the lapping process presented in Figure 3.7(b) in Section 3.7. The visible scratches on the surface of the wafer is due to the shear in a lapping process. These experimental observations of wafer surfaces provide more insights into the modeling of the process.

In summary, the modeling of the wiresaw manufacturing process includes the following objectives:

- To develop fundamental understanding of the complicated and multi-faceted manufacturing process.
- To avoid ad hoc control of process parameters in manufacturing.
- To reduce kerf loss.
- To achieve better control of process parameters (such as tension, speed, slurry composition, wire wear, feed rate, bow angle, etc.).
- To enhance surface quality of the sliced wafers through better understanding of the process.
- To develop on-line and real-time metrology of wire wear and control.

Figure 5.2 Typical surface of an as-sliced silicon wafer made by modern slurry wiresaw. The surface consists of homogeneous but random pits indented by abrasive grits (at 500X magnification). Source: Imin Kao.

• To provide knowledge in the design of reconfigurable wiresaws for customized operations in the future.

5.2 The Rolling-indenting Model

The fundamental cutting process of a modern wiresaw is described by the "rolling-indenting" process model, and is illustrated in Figure 5.3.

The free abrasive machining (FAM) process is associated with three-body abrasion, as presented in Section 3.4.2. In the FAM process prevalent in slicing using slurry wiresaws, an abrasive carrying slurry is brought by a wire web into the cutting zone (cf. Figures 4.7 and 4.8). During the FAM process of wiresawing, the cutting action is caused by abrasive grits carried by the slurry, which are trapped between the axially moving taut wire (the cutting tool) and the ingot being sliced (the workpiece), respectively. A force perpendicular to the direction of the moving wire is required to press the cutting tool onto the workpiece to remove material from the ingot surface to produce slices of wafers. The normal force is provided through a bow angle, α, of the wire spanning over the ingot. The bow angle is maintained at about $\alpha = 2°–6°$ to ensure a steady-state and uninterrupted progress of the cutting without wire breakage (cf. Figure 4.13).

Figure 5.3 provides a graphical illustration of the "rolling-indenting" mechanism, with an exaggerated cut-away view to show the interaction of abrasive grits in the rolling-indenting process. The wire moves at a very high speed ($V = 10–20$ m s^{-1}) with a film of viscous slurry trapped in between the wire and the stationary ingot surface. The boundary conditions of $V = 0$ at the ingot surface and the high-speed moving wire with $V = 10–20$ m s^{-1} cause the slurry film in between to assume a varying velocity profile from 0 to V, as shown in the figure. This velocity profile causes the finite-sized free abrasive grit trapped in slurry to have different speeds at the top and bottom of each individual abrasive grit, causing it to roll in clockwise direction, while carried by the viscous slurry from left to right in Figure 5.3.

At the same time, the abrasive grits are also acted upon by the hydrodynamic pressure of the slurry fluid [Bhagavat et al. (2000); Zhu and Kao (2005b)] (to be

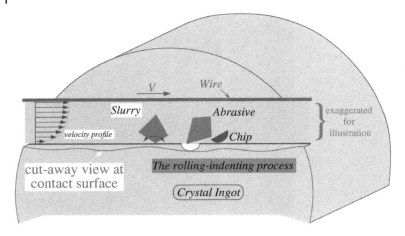

Figure 5.3 The engineering model of the slurry wiresaw manufacturing process: the "rolling-indenting" model. Note that the contact interface with the rolling-indenting interaction is exaggerated with a cut-away view for illustration.

discussed in Section 5.5), as well as the normal load caused by the bow angle of the wire, to press them to indent onto the surface of the ingot. Therefore, the abrasive particles will roll on the surface and at the same time be constrained to indent the surface – resulting in the so-called "rolling-indenting" process [Kao (2004); Li et al. (1998); Yang and Kao (2001); Kao et al. (1998a, b)]. Such interaction of rolling and indenting of abrasive grits causes materials to be removed from the surface of the brittle material through indentation with crack formation and propagation, such as silicon.

When abrasive grits indent onto the surface with loading and unloading actions, they cause the creation and propagation of radial and lateral cracks, as described in Section 3.3 and Figure 3.2(b). Figure 5.3 illustrates the results of the concatenation of subsurface cracks to render kerf and chips of small volume of material being removed from the surface of crystal materials. This results in the randomly distributed pits on the surface of wafers. The kerf and free abrasives are carried away by the slurry to exit the cutting zone. The slurry will then be collected in a slurry tank and pumped through the slurry manifold to be poured onto the wire web again.

Equations of stress due to contact and indentation are formulated with boundary conditions by Kao et al. (1997a,b) and Li et al. (1997, 1998). The results, expressed with dimensionless stress measures and normalized by the depth of indentation and angle of contact, suggest the regions of material removal for the ingot and the optimal shape of abrasives. It is recommended that the abrasive grits have a conical indenting angle between 90° and 120° to optimize the slicing process [Bhagavat et al. (1998)]. Maximum subsurface shear stress, characteristic of contact mechanics, facilitates the "peeling" effect and formation of chips on the ingot surface. The rolling-indenting model of slurry wiresaw describes a wire moving at very high speed in viscous slurry that is different from the slower lapping process in which

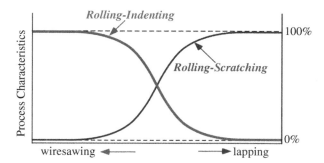

Figure 5.4 Illustration of the "rolling-indenting" and "scratching-indenting" mechanisms in FAM. The percentage of rolling-indenting is more prominent towards wiresaw, while that of scratching-indenting is more prevalent in lapping.

the relative speed between surfaces is much smaller, causing different interactions described by the scratching-indenting process.

In general, FAM processes, both rolling-indenting and scratching-indenting processes, can take place, as illustrated in Figure 5.4. The composition of both processes depends on the nature of FAM at the contact interface of machining.

5.3 Vibration Modeling and Analysis

A high-speed wire is employed in various wiresaw manufacturing processes, including slurry wiresaws and diamond-impregnated wiresaws (see Chapter 6); therefore, the understanding of vibration of a moving wire is important in improving the quality and yield of wafers and in minimizing kerf loss (see Equation 4.2). In this section, the vibration of a moving wire is analyzed under different operating parameters, such as tension and speed of the wire. Frequency response is used to examine the amplitude of vibration under a spectrum of excitation frequencies.

The wire employed in wiresaws is characterized by (i) low mass density[1], and (ii) high tension [Chung and Kao (2011, 2012); Wei and Kao (2000b)]. These two important features result in a very high critical speed for stability (see also Section 5.3.5), making the system stable under typical system parameters of operation. The excitation of vibration is a result of reaction forces exerted by abrasive particles and slurry interacting with the wire through the "rolling-indenting" process [Bhagavat et al. (2000); Bhagavat and Kao (2006); Kao (2002, 2004); Kao and Bhagavat (2007); Kao et al. (2010); Li et al. (1998); Yang and Kao (1999, 2001); Zhu and Kao (2005b)].

The presentation of vibration analysis in this section includes the following. First, a historical perspective of vibration of wire is presented. Next, the equation of motion

---

1 A wiresaw operates with a very thin steel cold-drawn wire, typically 175 μm or smaller in diameter, resulting in low mass per unit length. The mass density ranges typically from 0.15 g m^{-1} to 0.20 g m^{-1}. The tension of the wire ranges from 20 N to 30 N [Chung and Kao (2011, 2012); Wei and Kao (2000b)].

of a moving wire under the prescribed boundary conditions with the constraints of the wiresaw manufacturing process is derived. A non-dimensionalized equation of motion and boundary conditions is presented to facilitate the vibration analysis. After that, the modal analysis of the undamped system with a closed-form solution is derived, subject to harmonic excitations. Numerical results using Galerkin's method are compared with those of the closed-form solutions. The results of the frequency response with different wire speeds and tensions demonstrate that the tension of the wire has a much more pronounced effect on the amplitudes of vibration than the speed of the wire. These results suggest important strategies in the design and operation of wiresaws. Finally, recent advance in the analysis of damped vibration of a moving wire is presented. Damping is an inherited characteristic of a wiresaw system due to the damping effect of slurry fluid, which influences the amplitudes of vibration.

5.3.1   A Historical Perspective on the Vibration of Wire

Although the vibration of a stationary wire or string is an age-old problem and has been studied and solved since the 17th century, the vibration analysis of a moving wire is a very new research topic. The applications of this research can be found in diverse mechanical systems such as high-speed magnetic and paper tapes, high-speed fiber winding, thread lines, band saws, pipes that contain flowing fluid, power transmission chains, belts, and cable-driven elevators and equipment. The equation of motion of this continuous system was first derived by Archibald and Emslie (1958) and Swope and Ames (1963). The differential eigenvalue problem together with the modal analysis of vibration was first reported in the 1990s. A review can be found in a reference by Wickert and Mote (1988). The vibration of wire is modeled as the transverse vibration of an axially moving string under tension. An analysis of a moving wire in a undamped system was first reported by Wickert and Mote (1991) to find the eigenfuctions and natural modes of vibration. Recent research has been extended to the free vibration of an axially accelerating string [Pakdemirli et al. (1994)], a damped string vibration system [Huang and Mote (1995); Wei and Kao (2000b)], and active vibration control [Ying and Tan (1996)]. An extensive review of recent research on vibration and control of a moving wire can be found in [Chen (2005)].

The first closed-form solution of the eigenvalue problem of the damped vibration of a moving wire using the Green's function to produce an analytical and complete solution of vibration with damping was first published by Chung and Kao (2012). While the analysis of vibration of the moving wire in a damped environment, such as that in the wiresawing process, was first solved and reported by Chung and Kao (2011, 2012), other related research was presented in [van Horssen (2003a); van Horssen and Ponomareva (2005a); Wei and Kao (2000b); Zhu and Kao (2005b); Kao et al. (1998b)]. Ishikawa and Suwabe (1987) showed that the vibration in the cutting direction can increase the cutting speed in the wiresawing process. However, the surface quality was not reported. Although the vibration amplitudes of the wire are small [Chung and Kao (2011); Mizuno et al. (2006)], the requirements for surface quality of semiconductor wafers are very stringent (on the micron or nanometer scale). Therefore, the small amplitudes of vibration can have a profound effect on the quality of wafer surfaces.

5.3.2 Equation of Motion of a Moving Wire

Figure 5.5 illustrates the schematic of a moving wire in a modern wiresaw system, in which an elastic wire of a linear speed of V is carried by rotating wire guides with an angular speed of Ω_w, separated by a distance of L. Archibald and Emslie (1958) employed the Hamilton's principle to derive the equation of motion of a moving wire, while Swope and Ames (1963) used Newtonian mechanics. In summary, the partial differential equation of motion that governs a moving wire continuum is

$$\rho(U_{TT} + 2VU_{XT} + V^2U_{XX}) - PU_{XX} = F \tag{5.1}$$

with the boundary conditions based on the schematic in Figure 5.5

$$U(0, T) = 0 \quad \text{and} \quad U(L, T) = 0 \tag{5.2}$$

where U is the transverse displacement of the wire in the Y direction, V is longitudinal speed of the wire, P is the tension of the wire, F is the external excitation force applied on the wire per unit length, and ρ is the mass of wire per unit length (or the mass density).

The following dimensionless parameters are introduced to formulate Equation (5.1) to a non-dimensionalized equation of motion, as follows.

$$\begin{cases} x = & X/L \\ u = & U/L \\ v = & V\sqrt{\rho/P} \\ t = & T\sqrt{P/(\rho L^2)} \\ f = & FL/P \end{cases} \tag{5.3}$$

With the parameters defined in Equation (5.3), Equation (5.1) can be re-written as

$$u_{tt} + 2vu_{xt} - (1 - v^2)u_{xx} = f. \tag{5.4}$$

The boundary conditions in Equation (5.2) becomes

$$u(0, t) = 0 \quad \text{and} \quad u(1, t) = 0. \tag{5.5}$$

Figure 5.5 A schematic showing the wire and process parameters of a wire moving axially with a constant speed V in a typical wiresaw with rotating wire guides.

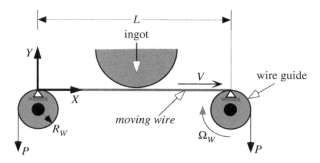

Note that the symbols in the upper case are physical parameters of the system, whereas symbols in lower case denote the non-dimensionalized parameters, defined in Equation (5.3). In the subsequent analysis, non-dimensionalized equation of motion in Equation (5.4) will be adopted for analysis.

5.3.3 Modal Analysis of an Undamped Moving Wire

Wickert and Mote (1990, 1991) applied the method of modal analysis on a undamped, axially moving wire and obtained a general closed-form solution for arbitrary initial conditions and excitations. The general closed-form solution of a undamped moving wire is an infinite series of the normal modes multiplied by time-dependent generalized coordinates [Wickert and Mote (1990)], as follows

$$u(x, t) = \sum_{n=1}^{\infty} \frac{\xi_n(t)}{n\pi \sqrt{1 - v^2}} \sin(n\pi x) \, \mathrm{e}^{(in\pi vx)} \tag{5.6}$$

where n denotes the mode numbers, $i = \sqrt{-1}$, x and v are the non-dimensionalized spatial coordinates and speed, respectively, as defined in Equation (5.3), and $\xi_n(t)$ are the complex generalized coordinates. Note that the response involves complex eigenfunctions subject to the boundary conditions in Equation (5.5).

5.3.4 Response for Point-wise Harmonic Excitation

Let's assume a single excitation at $x = x_0$ with a point-wise harmonic excitation function[2]

$$f(x, t) = f_0 \, \delta(x - x_0) \sin(\omega t) \tag{5.7}$$

where f_0 is the amplitude of excitation, x_0 is the location of excitation with frequency of ω, and δ is the Dirac delta function. All parameters are dimensionless.

Applying the modal analysis, the steady-state response for this point-wise harmonic excitation can be obtained as

$$u(x, t) = f_0 \sum_{n=1}^{\infty} \frac{1}{n\pi} \frac{1}{\omega_n^2 - \omega^2} \sin(n\pi x_0) \sin(n\pi x)[(\omega_n + \omega) \sin(\omega t + n\pi v(x - x_0))$$
$$+ (\omega_n - \omega) \sin(\omega t - n\pi v(x - x_0)) - 2\omega \sin(\omega_n t + n\pi v(x - x_0))] \tag{5.8}$$

where the dimensionless natural frequencies of the moving wire are

$$\omega_n = n\pi(1 - v^2) \tag{5.9}$$

with $n = 1, 2, 3, \cdots$ [Wei and Kao (2000b)]. The index n denotes the modes of eigenvalues and eigenfunctions of the forced vibration response. When the index $n = 1$, the frequency ω_1 is the fundamental natural frequency of the system.

The response of the forced vibration in Equation (5.8) is plotted in a 3D figure shown in Figure 5.6(left). The 3D graph shown in Figure 5.6(left) is plotted with

---

2 Such excitation can be caused by an abrasive particle in contact with the wire, for example. The assumption of harmonic excitation is a generic approach to derive the response of vibration.

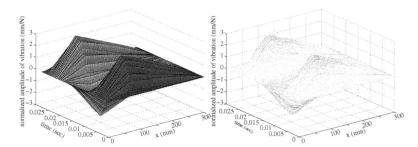

Figure 5.6 Three-dimensional plots of the normalized vibration amplitude under a harmonic excitation of $\omega = 0.1\omega_1$ (ω_1 is the fundamental frequency) at $x_0 = 75$ mm ($L = 300$ mm) with (left) apply the closed-form solution in Equation (5.8) (cf. Section 5.3.4), and (right) use the numerical solution by Galerkin's method (cf. Section 5.3.6)

the following system parameters: $V = 10$ m s^{-1}, $\rho = 0.1876$ g m^{-1}, $P = 20$ N, $L = 0.3$ m, $X_0 = 0.075$ m, $\omega = 0.1\omega_1$ and $n = 100$. The fundamental natural frequency is ω_1; thus, the frequency of external harmonic excitation is 10% of the value of the fundamental natural frequency in the simulation. The vibration is plotted across the span of the wire, with $0 \leq X \leq 300$ mm, as a function of time.

It is noted in Figure 5.6(left) that the excitation at the location of $X = 75$ mm causes the peaks of vibration to shift towards one side. In addition, the vibration pattern contains twin peaks and valleys because the response contains two main frequencies – one is the excitation frequency, ω, and the other is the fundamental frequency of the wire, ω_1. The two frequencies juxtapose with each other to render such a typical response. Refer to Wei and Kao (2000b) for a similar plot with the point-wise excitation at the middle of the wire span.

5.3.5 Natural Frequency of Vibration and Stability

Based on the definition of the dimensionless variables in Equation (5.3), the corresponding physical natural frequencies can be found from the non-dimensionalized natural frequencies that are derived in Equation (5.9) [Wei and Kao (2000b)]

$$\Omega_n = \omega_n \sqrt{\frac{P}{\rho L^2}} = \frac{n\pi(P - \rho V^2)}{L\sqrt{\rho P}}. \tag{5.10}$$

It is noted that when the speed of the wire, V, is zero, the natural frequency, Ω_n, in Equation (5.10) will reduce to the usual form of the natural frequencies of a stationary string

$$\Omega_n = n\pi \sqrt{\frac{P}{\rho L^2}}.$$

It is also obvious that the natural frequency Ω_n in Equation (5.10) will vanish when

$$V_{cr} = \sqrt{\frac{P}{\rho}}. \tag{5.11}$$

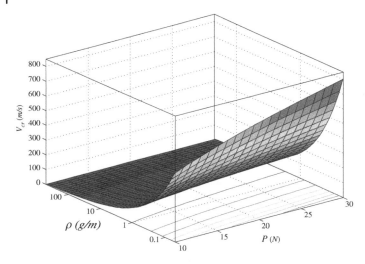

Figure 5.7 Three-dimensional surface plots of the critical speed of a moving wire, as a function of the mass density, ρ in g m^{-1}, and the tension, P in Newtons.

Equation (5.11) defines the *critical speed for stability* of a moving wire. Figure 5.7 plots the surface of the critical speed of a moving wire, V_{cr}, as a function of the mass density, ρ, in g m^{-1}, and the tension, P, in Newtons. The ranges of the plot are $0.0468 \leq \rho \leq 468$ g m^{-1} and $10 \leq P \leq 30$ N. Note that the mass density, ρ, is plotted in logarithmic scale for the purpose of illustration. The contour plots show as lines on the ρ–P plane, indicating constant V_{cr} on the surface. Equation (5.11) describes the critical speed as being proportional to the square root of tension and inversely proportional to the square root of mass density. The contour plots suggest that the tension and the mass density must increase at the same rate in order to keep V_{cr} the same. It is easy to observe that the values of V_{cr} decrease dramatically as the values of ρ increase.

Table 5.1 lists critical speeds and fundamental frequencies associated with wires of different diameters from 0.1 mm to 10 mm. As tabulated in Table 5.1, when the diameter of a steel wire is 1 mm, the mass density reaches 6.2 g m^{-1}, making $V_{cr} = 56.8$ m s^{-1}. This means that a 1 mm steel wire moving at a speed close to this V_{cr} will cause instability of vibration. Such instability can be remedied by increasing the tension of the wire.

Likewise, in the operation of a wiresaw system, the system of the axially moving wire will experience divergence instability when the speed of the wire is near this critical speed. For industrial wiresaws, the mass density, ρ, is very small while the tension, P, is high which results in a very high critical speed (typically over 300 m s^{-1}). The operating speed of the moving wire in a wiresaw system is normally less than 20 m s^{-1}; hence, the divergence instability is generally not a concern for the operation of wiresaws. An example with discussion is provided in the following.

Example 5.3.1

Determine the fundamental natural frequency of a moving wire in a wiresaw equipment. The following parameters are prescribed: the speed of the wire is $V = 10 \text{ m s}^{-1}$, the mass density of the steel wire, with a diameter of 175 μm, is $\rho = 0.1876 \text{ g m}^{-1}$, the tension of the wire is maintained at $P = 20 \text{ N}$, and the length of the wire between two supports at the boundary is $L = 0.3 \text{ m}$. Determine the fundamental natural frequency of the axially moving wire, as well as the critical speed of the wire for stability.

Solution:

By the definition in Equation (5.3), the dimensionless speed of the wire is

$$v = V\sqrt{\frac{\rho}{P}} = 10\sqrt{\frac{0.1876 \times 10^{-3}}{20}} = 0.3063.$$

Equation (5.9) can be used to determine the dimensionless fundamental natural frequency when $n = 1$.

$$\omega_1 = \pi(1 - v^2) = \pi\left(1 - V^2\frac{\rho}{P}\right) = \pi\left(1 - 100 \times \frac{0.1876 \times 10^{-3}}{20}\right) = 3.1387.$$

By substituting the given physical parameters into Equation (5.10), the fundamental natural frequency (when $n = 1$) of the moving wire is found to be

$$\Omega_1 = \frac{\pi(P - \rho V^2)}{L\sqrt{\rho P}} = \frac{\pi(20 - 0.1876 \times 10^{-3} \times (10)^2)}{0.3\sqrt{0.1876 \times 10^{-3}(20)}} = 3416 \text{ rad s}^{-1} \cong 544 \text{ Hz}.$$

Note that the fundamental frequency of a stationary wire with the same parameters is 3419 rad s^{-1}. The critical speed for stability is derived in Equation (5.11),

$$V_{\text{cr}} = \sqrt{\frac{P}{\rho}} = \sqrt{\frac{20}{0.1876 \times 10^{-3}}} = 326.5 \text{ m s}^{-1}.$$

The operating speed of wire in a wiresaw is much lower than the V_{cr} obtained above; thus, the instability of vibration is not a concern at all. It is also noted that the dimensionless speed can be found by

$$v = \frac{V}{V_{\text{cr}}} = \frac{10}{326.5} = 0.03063.$$

Example 5.3.1 provides a general idea of practical values of the dimensionless parameter of speed v, and the physical fundamental frequency of a moving wire.

Discussion of the Wire Size Versus the Critical Speed

For a homogeneous wire of diameter, d, with a cross-sectional area, $A_{\text{c}} = \frac{\pi}{4}d^2$, the mass density per unit length of the wire is

$$\rho = \rho_{\text{w}} A_{\text{c}} = \frac{\pi}{4} \rho_{\text{w}} d^2 \tag{5.12}$$

where ρ_{w} is the density of the material of the wire. For example, the density of steel is from 7750 to 8050 kg m^{-3}. Using Equation (5.12), the critical speed for stability and the fundamental natural frequency of a steel wire can be determined when operating under the following prescribed condition:

Table 5.1 Wire size versus the critical speed and the fundamental natural frequency; the speed, tension, and length of the wire are 10 m s^{-1}, 20 N, and 0.5 m, respectively.

Wire diameter (mm)	Mass density, ρ (g m^{-1})	Critical speed, V_{cr} (m s^{-1})	Ω_1 (rad s^{-1})
0.100	0.0620	568	3566
0.175	0.190	324	2037
0.50	1.551	114	208
1.0	6.205	56.8	113
5.0	115.1	11.4	16
10.0	620.5	5.68	–

- The speed of the wire is $V = 10$ m s^{-1}
- The volumetric density of the steel wire is $\rho_w = 7900$ kg m^{-3}
- The tension of the wire is maintained at $P = 20$ N
- The length of the wire between two supports at the boundary is $L = 0.5$ m.

Table 5.1 tabulates the results of critical speed of vibration and fundamental natural frequency with different diameters of a steel wire. Note that the mass density of the wire is calculated using Equation (5.12).

By comparing the example and Table 5.1, it can be deduced that a longer wire length and span will reduce the natural frequency, as expected, but does not affect the critical speed for stability. With constant speed and tension, the critical speed of stability and fundamental frequency decrease with larger wire diameters. When the tension is maintained at 20 N for a wire of 5 mm diameter, the critical speed for stability is 11.4 m s^{-1}, which is very close to the speed of the wire at 10 m s^{-1}. This will cause instability of vibration, unless the tension of the wire is increased in order to prevent the divergence instability, based on Equation (5.11). This makes sense intuitively because a heavy wire with small tension will become sloppy and floppy and will exhibit divergence instability. When the tension is increased to make the wire more taut, the instability can be prevented. When the wire diameter is increased to 10 mm, the target operating wire speed is larger than the critical speed for stability; thus, the wire is not operable, unless the tension of the wire is increased.

5.3.6 Numerical Solution Using Galerkin's Method

Equation (5.4) is a partial differential equation that can be discretized using Galerkin's method. The eigenfunction of a stationary string, $\sin n\pi x$, can be used as the trial function for the numerical analysis. The reason for choosing this eigenfunction, instead of the eigenfunctions in Equation (5.6), can be found in Wei and Kao (2000b). Thus,

$$u(x, t) = \sum_{j=1}^{n} q_j(t) \sin(j\pi x). \tag{5.13}$$

Applying Galerkin's method to discretize Equation (5.4) into the simultaneous ordinary differential equations gives

$$\mathbf{M}\ddot{q} + \mathbf{C}\dot{q} + \mathbf{K}q = \mathbf{Q}. \tag{5.14}$$

The components of the matrix $\mathbf{M}, \mathbf{C}, \mathbf{K}$, and the column vector $\mathbf{Q}$ in Equation (5.14) are

$$m_{ij} = \begin{cases} 1/2 & \text{if } i = j \\ 0 & \text{if } i \neq j \end{cases}$$

$$c_{ij} = \begin{cases} 0 & \text{if } i+j = 2n \\ (\frac{4ij}{i^2-j^2})v & \text{if } i+j = 2n+1 \end{cases}$$

$$k_{ij} = \begin{cases} \frac{i^2\pi^2}{2}(1-v^2) & \text{if } i = j \\ 0 & \text{if } i \neq j \end{cases}$$

$$q_i = \int_0^1 f \sin(i\pi x)\mathrm{d}x$$

where $i, j = 1, 2, \cdots, n$. It is straightforward from the discretized components listed above that $\mathbf{M}$ is a real symmetric positive definite mass matrix, $\mathbf{C}$ is a real skew symmetric gyroscopic matrix, and $\mathbf{K}$ is a real symmetric matrix. The stiffness matrix $\mathbf{K}$ is positive definite as long as the the speed of the wire is smaller than the critical speed, $V_{\mathrm{cr}} = \sqrt{P/\rho}$, derived in Section 5.3.5. It should be noted that when $v = 0$, the matrix $\mathbf{C}$ will vanish, resulting in a simplified symmetric eigenvalue system. This shows that the axially moving wire system is essentially a gyroscopic system while the stationary string is not.

The solution of Equation (5.14) can be found by using the modal analysis [for example, in Meirovitch (1997)]. By setting the matrix order to 10, i.e. $n = 10$, the result of simulation can be obtained, and is plotted in Figure 5.6(right).

It is noted that both the plots of theoretical and numerical results in Figure 5.6 are nearly identical to each other. The difference in the slightly rounded peaks and valleys of the vibration response in Figure 5.6(right) is due to the following two reasons:

(i) The use of the eigenfunctions of a stationary string instead of those of an axially moving string
(ii) The numerical round-off by choosing the finite matrix order at $n = 10$.

5.3.7 Response of Multiple-point and Distributed Excitations

Based on the analysis of the single-point excitation discussed in Section 5.3.4, the forced vibration analysis can be extended to the response of distributed multiple excitations. It is assumed that the excitations are distributed over the span of the wire and are described by the following equation

$$\sum_{j=1}^{m} f_j = \sum_{j=1}^{m} f(x_j, t) = \sum_{j=1}^{m} f_{0j}\, \delta(x - x_j) \sin(\omega_{f_j} t) \tag{5.15}$$

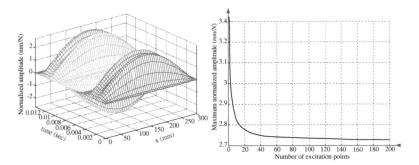

Figure 5.8 Left: Three-dimensional plot of the forced vibration analysis under distributed harmonic excitations of $\omega_f = 1024$ rad s^{-1}. A total of 100 excitations distributed between $x_0 = 50$ mm to 250 mm are applied. Right: Maximum normalized amplitude versus number of excitations. The amplitude asymptotically approaches a constant. Source: Wei and Kao (2000b). © 2000 Elsevier.

where m is the number of excitations, and f_{0j}, ω_{f_j} and x_j are the amplitude, frequency and location of the jth excitation, respectively. Each excitation in Equation (5.15) can be thought of as an abrasive grit making contact and applying a force on the wire at location x_j.

The system described in Equation (5.4) is linear if the system parameters, such as the longitudinal speed, mass density, and tension, are constants. Being a linear system, the response to the distributed multiple excitations in Equation (5.15) is the superposition of the response of each single excitation, derived in Section 5.3.4. Therefore, the response of vibration subject to distributed multiple excitations can be written as

$$u = \sum_{j=1}^{m} f_{0j} \sum_{n=1}^{\infty} \frac{1}{n\pi} \frac{1}{\omega_n^2 - \omega_{f_j}^2} \sin(n\pi x_j) \sin(n\pi x)[(\omega_n + \omega_{f_j}) \sin(\omega_{f_j} t + n\pi v(x - x_j))$$
$$+ (\omega_n - \omega_{f_j}) \sin(\omega_{f_j} t - n\pi v(x - x_j)) - 2\omega_{f_j} \sin(\omega_n t + n\pi v(x - x_j))]. \quad (5.16)$$

Figure 5.8(left) shows the three-dimensional plot of the response under distributed harmonic excitations with the same system parameters as those in Figure 5.6, under a total of 100 uniformly distributed excitations applied from $x_0 = 50$ mm to $x_0 = 250$ mm.

While 100 excitations are used to plot the response in Figure 5.8(left), it is of interests for us to know if the maximum normalized amplitude will vary under different number of excitations. Figure 5.8(right) presents the simulation results of the maximum normalized amplitude ranging from 1 to 200 excitations. The figure shows an immediate and appreciable drop-off in the amplitude when the total number of excitations is increased. As the number of excitation keeps increasing, the amplitude asymptotically approaches a constant. From Figure 5.8(right), it is observed that the maximum normalized amplitude does not change significantly when the number of excitation points exceeds 50. Therefore, a simulation of 100 excitations in Figure 5.8(left) is sufficient to capture the general response of a distributed vibration system of multiple excitations.

The numerical method in Section 5.3.6 can be employed with the principle of superposition to find the response of vibration under multiple excitations. The results are quite similar to those of the closed-form solution shown in Figure 5.8(left) [Wei and Kao (2000b)].

5.3.8 Frequency Response of Multiple Excitations

Frequency response is a methodology to determine the amplitude of the forced vibration response as a function of the frequencies of excitation. The maximum normalized amplitudes of vibration, subject to multiple excitations, are obtained for a range of excitation frequencies, under different tensions and speeds of the wire – the two important parameters of wiresaw operation. In the context of wiresaws, the following two questions are relevant to the analysis using the frequency response. The answers to these two questions and the outcomes of the analysis have profound implications on the design and operation of wiresaws.

1. What are the amplitudes of the vibration response of a moving wire under different tensions and speeds of the wire?
2. What is the most effective way to reduce the amplitudes of vibration, in order to reduce kerf loss? Speed or tension?

The closed-form response of forced vibration provided by Equation (5.16) is employed for the study of the frequency response of vibration. The excitation frequency is normalized with respect to the fundamental natural frequency, ω_1, of the moving wire. Without loss of generality, as discussed in Section 5.3.7, a total of 100 excitations is adopted in this analysis, uniformly distributed between $x_0 = 50$ mm and $x_0 = 250$ mm along the wire span with the same amplitude and frequency, ω. The amplitude of vibration is normalized with respect to the amplitude of excitation force. Parameter studies using the tension and speed of the wire are conducted by Equation (5.16), as well as the analysis in the preceding sections. The results are plotted in Figure 5.9.

Parameter Study Under a Constant Tension, $P = 20$ N

In this parameter study, the tension P is maintained at a constant value of 20 N, with the speed varying from 0 to 80 m s^{-1}. Figure 5.9(left) shows the frequency response obtained from Equation (5.16), where the maximum normalized amplitudes are plotted against the normalized excitation frequencies, ω/ω_1, with respect to different speeds of wire. The plot exhibits a discontinuity when $\omega/\omega_1 = 1$ where the amplitude becomes infinite due to resonance singularity. From the results in Figure 5.9(left), it is observed that the maximum normalized amplitudes are practically the same when the speed of wire is below 25 m s^{-1}. This is due to the small value of mass density, ρ, and high tension, P, which make the dimensionless speed trivial in responses when V is smaller than 25 m s^{-1}. For industrial wiresaws, the speed of wire is normally below 15 m s^{-1}; hence, it is concluded that the change in speed does not contribute much to the amplitude of vibration.

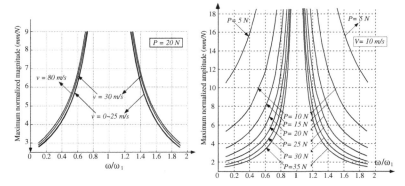

Figure 5.9 Left: Magnitudes of the frequency response with respect to different speeds of a moving wire. The speed of the wire varies from 0 to 80 m s^{-1} under a constant tension of 20 N. Right: Magnitudes of the frequency response with tension as a parameter. The tension of the wire varies from 5 N to 35 N at a constant wire speed of 10 m s^{-1}. Amplitudes of both figures are plotted with respect to the normalized frequency, ω/ω_1, where ω_1 is the fundamental natural frequency of the moving wire. Source: Wei and Kao (2000b). © 2000 Elsevier

Parameter Study Under a Constant Speed, $V = 10$ m s^{-1}

In this parameter study, the wire is kept at a constant speed of $V = 10$ m s^{-1}, with the tension varying from 5 N to 35 N. Again, Equation (5.16) is employed, with the results plotted in Figure 5.9(right). The results show that the maximum normalized vibration amplitudes are very sensitive to the variations of tension. When the tension is below 10N, the vibration amplitudes are very high. The higher is the tension, the smaller the amplitude of vibration. The figure also shows a non-linear effect of wire tensions on the amplitudes of vibration. As the tension becomes higher, further increase in tension does not result in a large reduction in the amplitudes of vibration. Therefore, it is not advantageous to increase the tension of wire indefinitely[3] for reducing the amplitudes of vibration.

Discussions

The results of the preceding analysis using the frequency response, as shown in Figures 5.9(left) and 5.9(right), suggest that a more effective way to reduce amplitudes of vibration and kerf loss of the wiresaw slicing process is to increase the tension of the wire, instead of the speed of the wire. As the process control and materials of wire improve, the tension employed in industrial wiresaws has been increased from 15N to 30N, resulting in considerable improvement of cutting efficiency and quality, as predicted by the results of the frequency response.

---

3 Practically, the maximum tension of the wire is subject to the yield strength of the material, and cannot be increased too much so as to risk wire rupture.

5.3.9 Vibration Responses of a Moving Wire with Damping

In the free abrasive machining (FAM) process of wiresawing, the high-speed wire moves through abrasive slurry between wafer surfaces (cf. Figure 5.5); thus, the slurry will cause a viscous damping effect on the wire. The damping force is assumed to be linear and is proportional to the speed of the wire. Therefore, the viscous damping force can be written as

$$F_\eta(X, T) = -\eta_d(VU_X + U_T) \tag{5.17}$$

where η_d is generally a linear constant, as will be discussed in Section 5.4, and $(VU_X + U_T)$ is the transverse velocity of the point on the wire located at X. Therefore, the equation of motion of a moving wire in Equation (5.1), now with damping, becomes

$$\rho(U_{TT} + 2VU_{XT} + V^2U_{XX}) - PU_{XX} = F + F_\eta \tag{5.18}$$

where F is an external distributed force excluding the damping forces. Equation (5.18) can also be normalized with dimensionless parameters introduced in Equation (5.3), with an additional dimensionless damping factor,

$$\eta = \frac{\eta_d L}{\sqrt{P\rho}}. \tag{5.19}$$

We can derive the non-dimensionalized equation of motion with damping, as follows

$$u_{tt} + 2vu_{xt} - (1 - v^2)u_{xx} + \eta vu_x + \eta u_t = f. \tag{5.20}$$

In general, a closed-form solution for such damped system can not be found [Meirovitch (1997)]. This remained the case until the analytical solution was presented by Chung and Kao (2011, 2012). Chung and Kao (2012) were the first to present the closed-form solutions for both free vibration and forced vibration responses using an innovative Green function for a moving wire continuum. The Green function for the moving wire continuum is $g(x, \xi, t - \tau)$, as follows

$$
\begin{aligned}
g(x, \xi, t - \tau) &= \sum_{n=1}^{\infty} \left[\frac{(1 - v^2)e^{(-\frac{\eta}{2} + i\omega_n)(t-\tau)}e^{\alpha_n(x-\xi)} \sin n\pi\xi \sin n\pi x}{i\omega_n} \right. \\
&\quad \left. - \frac{(1 - v^2)e^{(-\frac{\eta}{2} - i\omega_n)(t-\tau)}e^{-\alpha_n(x-\xi)} \sin n\pi\xi \sin n\pi x}{i\omega_n} \right] H(t - \tau) \\
&= \sum_{n=1}^{\infty} \frac{2(1 - v^2)e^{-\frac{\eta}{2}(t-\tau)}}{\omega_n} \left[\cos(\omega_n(t - \tau)) \sin\left(\frac{v\omega_n}{1 - v^2}(x - \xi) \right) \right. \\
&\quad \left. + \sin(\omega_n(t - \tau)) \cos\left(\frac{v\omega_n}{1 - v^2}(x - \xi) \right) \right] \sin n\pi x \, \sin n\pi\xi \, H(t - \tau)
\end{aligned}
\tag{5.21}
$$

where $\alpha_n = iv\omega_n/(1 - v^2)$ and $H(t - \tau)$ is the Heaviside function. The natural frequencies are

$$\omega_n = \frac{\sqrt{[4n^2\pi^2(1 - v^2) - \eta^2](1 - v^2)}}{2}.$$

The free vibration response was also presented in Chung and Kao (2011) for reference. For the sake of brevity, only the closed-form solution using the Green function is presented here. The readers may want to to refer to Chung and Kao (2011, 2012) for further reference on this advanced subject and the responses of vibration with plots.

The forced vibration response can be obtained by using the convolution integral of the Green function, $g(x, \xi, t - \tau)$, and external force, $f(x, t)$. The general formulation of vibration response using the Green function is as follows [Butkovskiy (1983); Roach (1982)]

$$u(x,t) = \int_0^t \int_0^1 g(x, \xi, t - \tau) f(\xi, \tau) \mathrm{d}\xi \mathrm{d}\tau. \tag{5.22}$$

The non-homogeneous initial and boundary conditions are carried into $f_I(x, t)$ and $f_B(x, t)$, respectively [Chung and Kao (2012)], to obtain the free and forced vibration responses. The solution of the vibration response, $u(x, t)$, is

$$u(x,t) = \int_{t_0}^t \int_D g(x, \xi, t - \tau)[f(\xi, \tau) + f_I(\xi, \tau) + f_B(\xi, \tau)] \mathrm{d}\xi \mathrm{d}\tau \tag{5.23}$$

where D is the domain of x and ξ. The term of external force, $f(\xi, \tau)$, is replaced by $f(\xi, \tau) + f_I(\xi, \tau) + f_B(\xi, \tau)$ in Equation (5.23).

5.3.10 Discussions

The vibration analysis in FAM of the wiresaw manufacturing process is presented systematically in this section. In the forced vibration analysis, the free abrasive grits in the slurry are treated as excitations to the wire to cause vibrations. After the presentation of historical perspectives, analysis of the moving string system and the equation of motion are presented. A non-dimensionalized equation of motion was derived with a set of dimensionless parameters, defined to facilitate the analysis. The solutions of free and forced transverse vibration responses of an axially moving wire are presented with modal analysis. The following list represents the progression of presentation:

- Equation of motion of an axially moving string continuum
- Modal analysis of a undamped moving wire
- Natural frequency and stability of vibration
- The free vibration response, as well as forced vibration responses with single and multiple excitations
- Frequency response and parameter study of tension versus speed of wire
- Numerical analysis using Galerkin's method
- The free and forced vibration responses of a damped moving wire, with the closed-form solution using the Green function.

The forced vibration results of single and multiple excitations are presented with both closed-form and numerical results. Both the closed-form and numerical results are shown to match each other closely. When the vibration analysis is based on multiple excitations, it is found that the maximum amplitude of vibration will asymptotically approach a steady-state value once the number of excitations reaches 50 or more. In a practical slicing process using a slurry wiresaw, there will be more than 50 abrasive particles making contact with the surface at any given time typically. This result is significant because it suggests that a steady-state amplitude of vibration can be used in the analysis of vibration for the purpose of predicting the kerf loss. For small vibration, the equations of motion in the horizontal direction and the vertical direction are the same; thus, the results are equally applicable to the vibration in both directions. While the vibration in horizontal direction will result in kerf loss, the vibration in vertical direction facilitates the "rolling-indenting" process, and hence the removal of materials.

In order to reduce the kerf loss, the amplitude of vibration of the wiresaw manufacturing process must be minimized across a spectrum of excitation frequencies. The analysis of frequency response reveals that it is more effective to increase the tension of wire for reducing the vibration amplitude and kerf loss than to change the speed of wire. When the tension is increased, the stiffness of the system is also increased – resulting in a reduction in the kerf loss [Sahoo et al. (1998); Wei and Kao (2004)]. It is important in the design of wiresaw equipment to realize that the amplitude of vibration, and hence the kerf loss, is not sensitive to the changes in wire speed, especially when the speed is below $25 \, \text{m s}^{-1}$. However, the increase in wire speed will affect other outcomes of manufacturing, such as materials removal.

When an axially moving wire is submerged in slurry, a viscous damping effect will be present in the system. The analysis of the damped vibration of a moving wire is presented with a recently developed closed-form solution using the Green function, applied to both free and forced vibration analyses. Due to the hydrodynamic effect of slurry in the wiresaw cutting process, it is desirable to have slurry that is neither too viscous nor too diluted [Bhagavat et al. (2000); Chung and Kao (2011, 2012); Kao et al. (1998a)]. A more complete presentation of this advanced topic of vibration analysis can be found in Chung and Kao (2011, 2012).

More references in vibration of a moving wire can be found, in addition to the ones referenced in this section, in the following: Archibald and Emslie (1958); Chen (2005); Chung and Kao (2008, 2011, 2012); Huang and Mote (1995); Ishikawa and Suwabe (1987); Ma et al. (2010); Malookani and van Horssen (2015); Oz and Pakdemirli (1999); Pakdemirli and Ulsoy (1997); Pakdemirli et al. (1994); Ponomareva and van Horssen (2009); Swope and Ames (1963); van Horssen (2003a,b); van Horssen and Ponomareva (2005a,b); Wei and Kao (1998, 2000a,b); Wickert (1994); Wickert and Mote (1988, 1990); Ying and Tan (1996); Zhu and Kao (2005b); Zhu et al. (2000); Kao et al. (1998b).

5.4 Damping Factor of the Slurry Wiresaw Systems

One very important, but often quite challenging, aspect of the vibration analysis of a damped system is the identification of realistic and practical damping factors. This is especially important for wiresaw systems with viscous slurries. Traditionally, there has been a lack of any previous data and analysis on the values of damping factors. The analysis of vibration with damping of a moving wire in wiresaw systems requires an accurate estimate of the physical damping factors. The following analysis presents an important contribution to the understanding of the damped vibration behavior of the moving wire in a slurry wiresaw machining tool [Chung and Kao (2011)].

Wiresaws are usually operated at very high speed (typically $10-15 \text{ m s}^{-1}$) and high tension ($20-35$ N). However, the actual non-dimensionalized speed of the moving wire in a wiresaw is far below the critical speed because of its low mass density (cf. Section 5.3.5). For example, the non-dimensionalized speed is $v = 0.0232-0.0459$ with typical process parameters of $V = 10-15 \text{ m s}^{-1}$, $\rho = 0.1876 \text{ g m}^{-1}$, $P = 20-35$ N, and $L = 0.2-0.5$ m. This value is much lower than the non-dimensionalized critical speed of $v = 1$. Therefore, we normally only have to consider for the analysis the situation when $v \ll 1$ in wiresawing operation.

For the flow passing through an immersed body, the force due to friction drag is more than the pressure difference when the Reynolds number is Re < 1. With this condition, the drag coefficient is $C_D \cong \frac{7}{\text{Re}}$ for a cylinder [Papanastasiou (1994)].[4] The Reynolds number can be formulated as

$$\text{Re} = \frac{\rho_f D V_{\text{tr}}}{\mu} \tag{5.24}$$

where ρ_f and μ are the density and viscosity of the fluid, D is the diameter of the wire, and $V_{\text{tr}} = V U_X + U_T$ is the transverse speed of the wire. Polyethylene glycol, a water-soluble carrier for slurry, has a density of $1.1-1.2 \text{ g cm}^{-3}$. The density of silicon carbide is 3.21 g cm^{-3}. Therefore, the density of the slurry is $\rho_f = 1491-1763$ kg m^{-3} for mixing ratios of $0.75-1.25$ kg of grit per liter of carrier. The viscosity is $\mu = 200-1000$ cP. In industry applications, different recipes to optimize the slicing process may be used, resulting in a slightly different range of parameters.

Since the slope of the wire, U_X, is small, the transverse speed of the wire is approximated as U_T, neglecting the term $V U_X$. The maximum U_T can be approximated as the product of vibration amplitude of the wire and natural frequency. The amplitude of vibration of the wire is assumed to be $25-150$ μm by subtracting the wire diameter, $150-175$ μm, from the kerf loss, $200-300$ μm. The natural frequency of the first component is $2047-6781 \text{ rad s}^{-1}$ with the following parameters

- The range of physical speed of the wire is $V = 10-15 \text{ m s}^{-1}$
- The mass density of the wire is $\rho = 0.1876 \text{ g m}^{-1}$
- The range of tension of the wire is $P = 20-35$ N
- The range of the length of the wire span between wire guides (cf. Figure 5.5 and Section 5.3.2) is $L = 0.2-0.5$ m.

Therefore, the transverse speed of the wire is $U_T = 0.0512-1.017 \text{ m s}^{-1}$. The true transverse speed should be smaller than this range due to damping. With these

---

4 This is not the only equation of the drag force in fluid flow for a cylinder at small Reynolds number. Other literature such as Sherman (1990) has a different expression. However, the results of analysis are within the same order of magnitude.

Figure 5.10 Free vibration response with with initial displacement 0.01 sin πx and different damping factors $\eta = 2, 2\pi$, and $\eta = 28$. The non-dimensionalized velocity v is kept at 0.033. The component n is from 1 to 10. The response at the middle of the wire becomes non-oscillatory when the damping factor $\eta \geq 2\pi$.

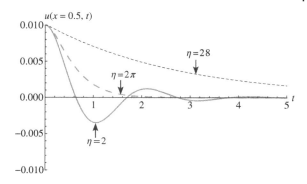

parameters, the Reynolds number in Equation (5.24) is found to be 0.0115–1.5688. Although the range is not always smaller than one, it is still small enough for the drag coefficient to be applicable. As a result, the coefficient of $C_D \cong \frac{7}{Re}$ is employed to evaluate the damping factor [Papanastasiou (1994)].

The function due to drag force is $f_D = C_D \frac{\rho_f V_{tr}^2 A}{2}$, where $A = DL$ is the projected area of a cylinder. Because $C_D \cong \frac{7}{Re}$ and $Re = \frac{\rho_f D V_{tr}}{\mu}$, the drag force for a cylinder is $f_D = 3.5\mu V_{tr} L$. The damping force per unit length then becomes

$$F_\eta = \frac{f_D}{L} \cong 3.5\mu V_{tr}.$$

It is also noted that $F_\eta = \eta_d V_{tr}$. Therefore, the damping factor, η_d, can be obtained as

$$\eta_d \cong 3.5\mu. \tag{5.25}$$

Here, the viscosity of slurry is $\mu = 200$–1000 cP [Kao (2004)]. Thus, the approximate range of the damping factor is $\eta_d = 0.7$–3.5 N s m^{-2}. The corresponding nondimensionalized damping factor is $\eta = 1.728$–28.57. Figure 5.10 illustrates the vibration response at the middle point of wire with initial displacements of $a(x) = 0.01 \sin \pi x$ and no initial transverse velocity [Chung and Kao (2011, 2012)]. The non-dimensionalized velocity and damping factors correspond to the wiresaw system discussed earlier with $v = 0.0306$, $\eta = 1.728$–28.57, respectively. A non-oscillatory response is also plotted, as shown in Figure 5.10 with $\eta = 2\pi$, by setting the frequency of the first mode equal to zero, $\omega_{1,d} = 0$. This gives damping factor $\eta = 2\pi\sqrt{1 - v^2} \approx 2\pi$ with $v = 0.0306$. As illustrated in Figure 5.10, non-oscillatory responses occur when $\eta \geq 2\pi$.

5.5 Elasto-hydrodynamic Process Modeling

The FAM process is associated with three-body abrasion, as presented in Section 3.4.2. In the FAM process prevalent in wiresaw slicing, a fraction of an abrasive-carrying slurry is brought by a wire web into the cutting zone (cf. Figures 4.7 and 4.8). During the FAM process of wiresawing, the cutting action is caused by fine abrasives freely dispersed in the slurry, which get trapped between the cutting tool and the workpiece[5]. A force perpendicular to the direction of cutting is required to provide the normal load and remove material from the ingot to produce slices of wafers. The normal load is supplied through a bow in the wire, which is maintained at about $\alpha = 2°-6°$ for an uninterrupted progress of the cutting process, without wire breakage (cf. Section 4.13).

The wiresaw FAM provides a unique example of how various constituents of a manufacturing system act concurrently to form an elasto-hydrodynamic (EHD) environment. Other examples of the same phenomenon are the traditional wire drawing and hydrostatic extrusion processes. Elasto-hydrodynamic interactions are found in such processes, even with a light lubricant. Journal or roller bearing is another prominent example of the elasto-hydrodynamic phenomenon, which involves EHD lubrication. The Newtonian fluid, such as the lubricant in bearings or slurry in wiresaws, interacts with the sliding element and the supporting surface to increase the pressure and create a film between the surfaces with relative motion.

The slurry, coupled with the wire's specific axial profile, generates an environment conducive to the development of hydrodynamic pressure, and hence slurry film in this region. This EHD interaction is an important part of the physics of the slicing process using a slurry wiresaw. It explains the entry of abrasives into the cutting zone between the taut wire and the ingot, which is governed by EHD interaction between the slurry and the wire [Bhagavat et al. (2000)]. In addition, an EHD film is formed by the abrasive-carrying viscous slurry, squeezed between the wire and the ingot. This phenomenon was analyzed and presented using the finite element method, seeBhagavat et al. (2000) and Zhu and Kao (2005a,b). The analysis of such an interaction involves coupling of the basic Reynold's equation of hydrodynamics with the elasticity equation of wire. Such comprehensive analysis yields a film thickness profile and pressure distribution as a function of wire speed, slurry viscosity, and slicing conditions. The results of film thickness and pressure distribution suggests that the wiresawing occurs under "floating" machining condition. The minimum film thickness is greater than the average abrasive size. This is practically important since the wiresaw is used to slice brittle semiconductor wafers with stringent requirements on the surface finish.

5.5.1 Approach of Modeling of EHD in the Wiresawing Process

Conceptually, the situation is not very different from the very well-known case of low wrap angle flexible foil bearings [Ma (1965)]. The EHD interaction is modeled as a compliant three-dimensional slider bearing, with the ingot as rigid and the moving wire under tension as elastic. The slurry is approximated as a Newtonian fluid with the actual viscosity measured in characterization experiments. Based on the typical viscosity data presented in Figure 3.10 and Section 3.8, it can be seen that the actual behavior is not markedly different from the Newtonian behavior, at least for the high shear rates prevalent in thin laminar flows. This is valid for a slurry with low abrasive

---

5 The cutting tool is the axially moving taut wire and the workpiece is the ingot being sliced, respectively.

concentration, wherein the mean size of abrasives in the slurry is less than the slurry film thickness.

In slicing silicon ingots for electronic and photovoltaic applications, the typical size of the abrasives used is in the range of 10 to 25 μm, with a slurry concentration of about 30% [Kao et al. (1998a)]. The wire moving towards the ingot carries with it a layer of slurry due to surface tension. The effective thickness of this layer, which is seen to completely cover the wire web, can be taken as at least half the thickness of the wafer being cut.

Momentum principles have been used to determine the cutting zone inlet condition, wherein the impacting slurry jet generates a pressure head at the inlet to the narrow film passage. At the point of entry of the wire into the cutting zone, this slurry layer impacts onto the ingot. This impact pressure, although small, when coupled with the high compliance of the wire at the inlet, causes the spacing between the ingot and the wire to open. This generates the initial converging passage for the slurry to enter the cutting zone. The work on the pressure-head buildup at the inlet of narrow passages provides the basis for modeling this impact pressure head [Tipei (1978, 1982)].

Once the slurry enters the converging inlet to the cutting zone, hydrodynamic interaction begins. A converging passage, formed due to hydrodynamic pressure and compliant wire at the inlet, helps in developing hydrodynamic pressure. The highly viscous slurry generates a hydrodynamic pressure. The elastic taut wire deforms under this pressure, generating an EHD environment. The film thickness is governed by the equilibrium between the load exerted by the deformed wire and the hydrodynamics.

The assumptions of analysis presented in Section 5.5.2 include:

- The film thickness is very small as compared to the cut profile radius and as such the compliance of the taut wire is linear.
- Thin film formulation with no variation of pressure along the film thickness.
- The slurry behaves as a Newtonian fluid.

5.5.2 Theoretical Modeling

The film thickness in EHD depends on both the geometry-dependent nominal profile and the pressure profile. For two-dimensional thin films, this dependence can be expressed as [Oh and Huebner (1973); Rohde and Oh (1975, 1977a,b)]

$$h(x,y) = h_g(x,y) + \Lambda(x,y)\, p(x,y)\, f(y) + h_{ref} \tag{5.26}$$

where $\Lambda(x,y)$ is the compliance influence coefficient[6], the function $f(y)$ is a geometry dependent mapping factor[7], and h_{ref} is the slurry film reference thickness that is related to compliance of the wire. The primary function of h_{ref} is to overcome

---

6 The compliance influence coefficient is a linear operator relating pointwise deflections to global pressure distribution.

7 The *mapping factor* is used to map the existing three-dimensional situation to a two-dimensional field.

the ever dwindling sphere of attraction of the Newton's method with increase in load. At the same time, this reference film thickness can also be manipulated to achieve the correct load bearing characteristic. This in turn ensures that the algorithm is proceeding to the correct load bearing solution. The film thickness can be expressed as

$$h(x,y) = h_g(x,y) + \Gamma(x,y)\,p(x,y) + h_{\text{ref}} \tag{5.27}$$

where Γ is a modified compliance operator independent of pressure given by $\Gamma = \Lambda(x,y)f(y)$.

The basic Reynold's equation for a steady-state, two-dimensional, incompressible, hydrodynamic flow, with the flow in the x-direction (the horizontal direction in Figure 5.3) is [Gohar (1988)]

$$\Psi(p) = \frac{\partial}{\partial x}\left(\frac{h^3}{\mu}\frac{\partial p}{\partial x}\right) + \frac{\partial}{\partial y}\left(\frac{h^3}{\mu}\frac{\partial p}{\partial y}\right) - 6U\frac{\partial h}{\partial x} = 0. \tag{5.28}$$

The Newton–Raphson iteration procedure can be employed to solve this non-linear integro-differential equation.

Boundary Conditions

The Swift–Steiber (SS) boundary condition is applied for maximal load capacity [Cameron (1966)]. In addition, the SS boundary condition is easy to implement in numerical schemes to obtain the solution. The SS boundary condition states that at the inlet boundary

$$\frac{\partial p}{\partial n_n} = 0 \quad \text{and} \quad p = 0 \tag{5.29}$$

where n_n is the outward normal to the boundary. The first pressure condition is the natural boundary condition for the formulation of the problem. The second condition is incorporated in the FEM program using the penalty functions.

The same SS boundary conditions have been enforced on the exit side, but with a difference. In order to avoid the tensile stresses in the film, the boundary condition is applied at the first instance where the analysis yields negative pressures in the course of the iteration procedure. This actually establishes the outlet boundary.

The details of the numerical algorithm using the Newton–Raphson method can be found in Bhagavat et al. (2000).

5.5.3 Results of the EHD Analysis

Figure 5.11 plots the elasto-hydrodynamic pressure distribution across the axial direction of the wire. The cutting zone of the ingot spans 100 mm. The parameters of the slicing process used in the simulation are: slurry viscosity of 1000 cP, wire speed of $V = 10$ m s^{-1}, bow angle of $\alpha = 2°$, and ingot length of 100 mm. The unit of pressure is in MPa. Figure 5.12 plots the corresponding film thickness in mm across the axial direction of wire, using the same process parameters.

From Figure 5.11, the following observations are presented for the EHD pressure profile.

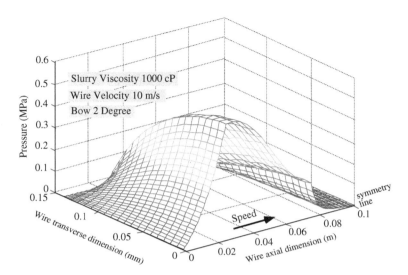

Figure 5.11 Elasto-hydrodynamic pressure profile in the cutting zone for 100 mm ingot length. Source: Bhagavat et al. (2000). ©2000 American Society of Mechanical Engineers.

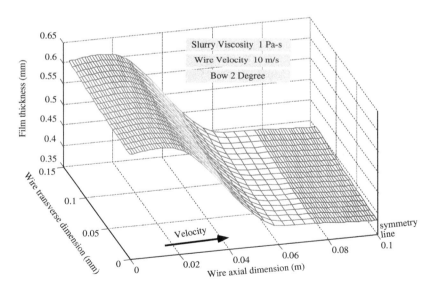

Figure 5.12 Profile for the film thickness in the cutting zone for a 100 mm ingot length. Source: Bhagavat et al. (2000). ©2000 American Society of Mechanical Engineers.

1. **At the inlet:** In the vicinity of the inlet to EHD zone, the pressure builds up gradually in the direction of the wire movement, shown by the arrow in Figure 5.11, consistent with the boundary conditions. The rate of increase of this pressure profile, as well as the peak pressure, depends on the slurry viscosity, wire speed, and bow angle of the wire. The maximum pressure is about 0.5 MPa, and is attained at about 30 mm downstream of the inlet, with the prescribed process parameters. The load to be sustained by the hydrodynamic action remains constant with the wire bow angle of $\alpha = 2°$, which remains constant under the constant tension in wire.

2. **At the outlet:** Towards the exit side, the EHD pressure starts decreasing from the peak value monotonically until a small spike is produced in the pressure profile. Finally, at around 85 mm from inlet the pressure drops down to nearly zero, thus marking the end of the hydrodynamic domain.

3. **Characteristics of EHD:** The pressure decline towards the exit side exhibits a plateau (or a spike) that appears as a temporary halt or increase in the pressure. The steep pressure gradient in the outlet region of the EHD domain would require a substantial increase in flow rate leading to severe flow continuity problems. To compensate for this, a small plateau of constant pressure, or temporary increase, results in the otherwise monotonically decreasing pressure profile. Such a phenomenon is quite common in majority of EHD situations, for example, in Hamrock (1994); Szeri (1980).

4. In the transverse direction, the pressure is constant except near the boundaries where it suddenly tapers down to zero due to the boundary condition.

Using the same process parameters, the hydrodynamic film thickness is plotted in Figure 5.12. It is observed from the figure that the thickness of the film is relatively large at the inlet. Thereafter, the thickness reduces downstream of the inlet, thus forming a converging passage for the pressure to build up. At about 20% of the cutting length after the inlet, the film thickness converges to a final and steady film thickness towards the exit. This steady film thickness is maintained as constant until the exit. The thickness along the transverse direction practically does not vary. This is because it was assumed that the wire cross-section remains circular throughout the deformation in compliance calculations.

The film thickness profile is quite contrary to what can be expected from the pressure profile. This unusual behavior can be explained from the compliance influence coefficients for the wire as it travels over a circular arc. The details of this explanation can be found in Bhagavat et al. (2000).

5.5.4 Implications Related to Floating Machining and Rolling-indenting Modeling of Modern Slurry Wiresaws

In a slurry wiresaw machining, a fraction of the slurry gets trapped in the cutting zone between the wire (tool) and the ingot (workpiece). This trapped slurry, driven by the wire moving over a near circular-arc cut profile of the ingot, undergoes a shear (Figure 5.3). This, coupled with the wire's peculiar axial profile, generates an environment conducive to the development of hydrodynamic pressure and hence slurry film in this region. The EHD pressure, along with the shear stress associated with the laminar velocity profile of the slurry, can cause the rolling and indenting of abrasive grits on the surface of workpiece, common to *three-body abrasion* phenomena, illustrated in Figure 5.3.

Based on the preceding analysis, the typical EHD film thickness is greater than the abrasive size in slurry wiresaw operations. This means that the abrasives are not directly pressed by the wire onto the ingot. Rather, the abrasives cause machining action in a "floating" state of rolling-indenting interaction, as presented in Section 5.2. Such a non-direct-contact abrasive machining mechanism is expected to be very inefficient from a material removal point of view. It is a known fact that the slurry wiresaw slicing process is slower than its counterpart (ID saw) for traversing the entire cross-section of an ingot. However, this is compensated in the industrial wiresaws by cutting 200–300 wafers at the same time, resulting in a higher overall throughput of wafer production.

Traditionally, "float" slicing and polishing processes with three-body abrasion have been used to manufacture components of fragile brittle materials with fine surface finish. Such a "float" slicing process has immediate advantages. Since abrasives are homogeneously distributed within the hydrodynamic film, a close to uniform distribution of abrasives in the cutting zone is formed. This ensures that the load sustained by individual abrasives is more or less uniform. The immediate manifestation of such an action can be observed in microscopic images of an as-sliced surface in Figure 5.13(a). This surface has an appearance composed of a number of uniformly and yet randomly distributed, and almost equal-sized, pits – a characteristic of individual abrasive action. Such a surface signature cannot be justified by a model involving abrasives directly pressed by the wire onto the ingot, since scratching and plastic deformation would be expected under this mode of material removal. As a comparison, a microscopic image of a lapped surface of silicon wafer is shown in Figure 5.13(b), after it is polished with a conventional lapping machine under lower speeds. A typical signature of a semi-dry three-body abrasion can be clearly seen. Such a surface is composed of a number of distinct scratches and pits caused by few abrasives bearing most of the load. In such polishing process, the abrasives are directly pressed onto the surface by the polishing wheel or pad. Only few abrasives of favorable size and orientations bear the polishing loads and take part in material removal. This gives strong indirect evidence of the role played by EHD in a typical wiresawing process with slurry.

Even with floating three-body abrasive machining modeling, pure rolling by itself cannot cause machining. This rolling tendency can be reduced by introducing an indenting force on the abrasive. Such an indenting force is provided by the EHD pressure, as presented earlier. Thus the EHD film and pressure, coupled with the inherent surface roughness of the ingot, causes rolling and indenting of abrasives that is the mechanism for manufacturing. Such a *rolling-indenting model* is very well known in the case of wet FAM processes [Bhagavat et al. (2005, 2010)]. See also Sections 3.7 and 5.2 for more detail.

The material removal rate by a single abrasive in the rolling-indenting process was estimated in Bhagavat et al. (2000) as a function of hydrodynamic parameters, and

(a) wiresaw-sliced wafer surface

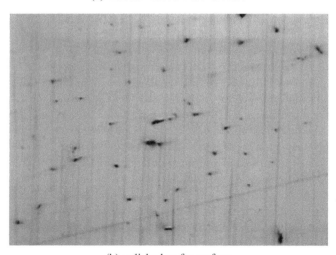

(b) polished wafer surface

Figure 5.13 (a) Surface of an as-sliced silicon wafer by the wiresaw shows a uniform and random distribution of almost equal-sized pits. Such phenomenon is justifiable only with floating abrasives. (b) The surface of a silicon wafer after it is polished at lower speeds shows distinct scratches and sparse pits. This indicates that the abrasives are directly pressed unto the ingot by the tool. Only those abrasives of favorable orientation and size will sustain load to cause polishing. Source: Bhagavat et al. (2000).

can be expressed as

$$R_{mr} \propto \mu \, a \, p \left(\frac{U}{h} \right)^2 \qquad (5.30)$$

where μ is the viscosity of the slurry, a is the average size of abrasives, p is the average hydrodynamic pressure, U is the wire speed, and h is the average hydrodynamic film thickness. As shown by Equation (5.30), the material removal rate of FAM is a complex function of hydrodynamic parameters. The material removal rate versus speed curves can be plotted in accordance with Equation (5.30) with the prescribed parameters. An example of such a plot of R_{mr} under constant viscosity, for an ingot of 50mm length with a 2° bow angle of wire was illustrated in Bhagavat et al. (2000). The results show that the material removal rate R_{mr} increases with speed and viscosity (and hence also with film thickness).

5.5.5 Important Conclusions from EHD Modeling

One important conclusion from this study suggests that the process of slicing ingot to wafers occurs through floating abrasives in typical wiresaw operations, known as the *rolling-indenting* process. This is well supported by the appearance of random pits on the surface of the as-sliced wafers. The abrasives usually are not directly pressed by the wire onto the ingot in a slurry wiresaw. Instead, the abrasives, under the influence of fluid shear generated by the slurry flowing at high velocity gradient, remove materials from the substrate surface through the rolling-indenting action. This is enforced by the hydrodynamic pressure. The material removal rate increases with wire speed and viscosity, as well as the film thickness.

There are situations in slicing operations during which abnormal cutting conditions occur, such as localized depletion of slurry or slurry dry out throughout the cutting zones. Under such conditions, direct contact among free abrasive grits, wire and ingot surface may take place that typically results in a poor surface with visible scratch lines on wafer surfaces or wire rupture due to excessive dry friction. In some cases, residual stress can be generated in wafers that renders them useless. These conditions are to be avoided in slurry wiresaw operations.

Another important conclusion is that the EHD effect does not percolate deep into the crystal. This can be a shortcoming for slicing crystal ingots of large diameter. Therefore, the to-and-fro reciprocating motion of the wire is necessary for efficient and uniform cutting of large-diameter crystals.

The wire bow, in conjunction with the abrasive size and slurry concentration, has a significant effect on the rate of material removal and productivity of the process. The hydrodynamic film thickness would be lower for a larger wire bow. If the film thickness is much lower compared to the abrasive size, then the abrasives would not enter the cutting zone, thus leading to very low slicing effectiveness with a risk of wire rupture due to dry friction with stick-slip. This is a primary reason for the use of very fine abrasives in the saws running at low speeds of 2–3 m s^{-1}. For such saws, it is prudent to maintain the wire bow at 2° or below. For modern slurry wiresaws with wire speeds higher than 10 m s^{-1}, the wire bow can be maintained at roughly 4–6° and larger abrasives can be used.

5.6 Thermal Management

A few degrees of fluctuation in the operating temperature of a slurry wiresaw can have a profound impact on the quality of the wafer surface. For example, a fluctuation in temperature of the wafer mount-and-support assembly and the wire guides spanning the wire web can cause a spatial waviness on the wafers being sliced. As an example of quantitative assessment, we use the following parameters:

- The wafer mount-and-support assembly and/or wire guides are made of steel which has a coefficient of thermal expansion $\alpha = 12 \times 10^{-6}$ m m^{-1} °C^{-1}.
- The temperature fluctuation is assumed at $\Delta T = 3$ °C.
- The length of the wafer mount-and-support assembly and/or wire guides is $L_0 = 0.5$ m.

The change in length due to $\Delta T = 3$ °C is

$$\Delta L = L_0 \, \alpha \, (\Delta T) = 0.5 \times (12 \times 10^{-6}) \times 3 = 18 \times 10^{-6} \text{ m} = 18 \text{ μm}. \tag{5.31}$$

Such variation of a few tens of microns, although small to typical bulk machine tools, is in the same order of magnitude as the slicing parameters and surface variation of wafers. A fluctuation of 18 microns in the wafer mount-and-support assembly and wire guides for a mere change of temperature of 3 °C is enough to cause spatial waviness on the surface of sliced wafers!

The preceding analysis brings out the relevance of thermal management in a slicing process using wiresaw equipment. Spatial waviness and warp can be generated as a direct result of temperature variation from rotating spindles and moving parts. Cooling mechanisms to such critical components of wiresaw become important when high and consistent wafer surface finish is required. In order to reduce surface waviness and warp in nanotopology, proper thermal management is necessary, for example, with cooling at influential parts of assembly of the wiresaw equipment. Several relevant patents [Gupta et al. (2011); Toyama et al. (1992); Zavattari et al. (2013)] are listed for further reference.

5.7 Wire, Wire Web, and Slurry Management

In the following sections, we will discuss the wire, wire web management and slurry management during wiresawing operations. The integrity of wires of the wiresaw machines is important to ensure that they will not rupture prematurely due to the wear of wire (reduction of diameter) during the slicing process. It is equally important to ensure that the wire web onto which the ingot is fed for slicing is maintained with the prescribed spacing without distortion throughout the slicing process, due to the wear of wire guides and grooves on which the wire seats. Two areas of real-time and on-line monitoring of the integrity of wire and wire web are described in the following sections. The slurry management is also discussed.

5.7.1 Real-time and On-line Monitoring of Wire Wear

During the operating of wiresaws, it is important to monitor the amount of the wire wear. If the wire becomes too thin after extended usage or display abnormal thickness distribution, it can break and cause significant downtime and waste of ingot[8]. Nonetheless, it is not advisable to check the wear of the wire by periodically stopping the operation. Thus, it is important that a real-time monitoring methodology be developed. In the following, a schematic of an optical setup is described to perform real-time and on-line monitoring of wire wear. The schematic of the system is shown in Figure 5.14.

As shown in Figure 5.14(a), a low power (e.g., 10 mW) He-Ne laser is used to illuminate the wire, which gives rise to a diffraction pattern that can be captured by a CCD sensor behind the ground glass screen. The separation between any two diffraction orders of a diffraction pattern is determined by

$$\Delta = \frac{\lambda L}{d} \tag{5.32}$$

where λ is the wavelength of the laser light, L the distance between the wire and the ground glass screen, and d is the diameter of wire that gives rise to the diffraction pattern, as illustrated in Figure 5.14(a). The separation between the diffraction orders are different in Figure 5.14(b) and 5.14(c), due to the difference in the size of the wires which are 175 µm and 146 µm, respectively. A linear CCD light sensor can be placed behind two neighboring diffracting orders whose output is used to calculate the separation of the orders, Δ, which in turn is used to calculate the wire diameter in real time, as follows, from Equation (5.32)

$$d = \frac{\lambda L}{\Delta}. \tag{5.33}$$

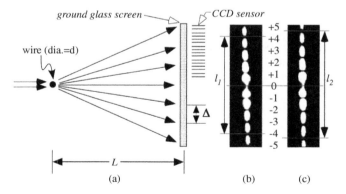

Figure 5.14 Optical arrangement for real-time monitoring of wire diameter during wiresaw operation: (a) optical system setup, (b) diffraction pattern from a 175 µm wire, and (c) diffraction pattern obtained from a 146µm-diameter used wire after a wiresawing operation.

---

8 If the wire ruptures in the middle of slicing, the aborted process will cause a step rise on the surface of wafers, even if a new wire is spun to form a new wire web for slicing. The entire crystal ingot should be discarded when this occurs.

Equation (5.33) stipulates that the diameter of the wire is inversely proportional to the separation of the orders, Δ. This is reflected in the separations of the orders of the patterns presented in in Figures 5.14(b) and 5.14(c) where the separation is larger for the wire with smaller diameter. By measuring the separation and calibrating to obtain accurate distance, the diameter of the wire can be obtained in real-time to monitor the wire wear using Equation (5.33).

Example 5.7.1
Determine the diameter of a moving wire in wiresaw equipment using the diffraction equipment in Figure 5.14. A He–Ne laser was employed with the length between the wire and the screen being $L = 0.6$ m. The separation of the orders of the patterns is found to be $\Delta = 2.4$ mm. Determine the diameter of the wire.

Solution:
The He–Ne laser has a wavelength of $\lambda = 632.8$ nm. Applying Equation (5.33), we find

$$d = \frac{\lambda L}{\Delta} = \frac{(632.8 \times 10^{-9})(0.6)}{2.4 \times 10^{-3}} = 158.2 \times 10^{-6} = 158.2 \ \mu m.$$

Thus, the diameter of the wire is 158.2 μm. If the initial diameter of wire is 175 μm, the wire wear is 9.6%. The equations presented here can be used to monitor the real-time diameter of the wire to ensure that it does not became dangerously thin to cause rupture.

Alternatively, the diameter of a wire with unknown diameter can be calibrated and calculated using the results of two such diffraction patterns, as those presented in Figures 5.14(b) and 5.14(c). Let's assume the linear distance between ±4 diffraction orders of the diffraction patterns of the 175 μm and 146 μm, as labeled in Figures 5.14(b) and 5.14(c), are l_1 and l_2, respectively, where $l_1 < l_2$. The separation between ±4 diffraction orders will be $l = 8\Delta$. Measuring l_1 and l_2 and dividing by 8 to obtain Δ_1 and Δ_2 will reduce the measuring error and increase the accuracy of the distance Δ. From Equation (5.33), we know that Δ is inversely proportional to wire diameter, d; therefore, we can write

$$d_1 \Delta_1 = d_2 \Delta_2 = d \ \Delta \Rightarrow d_1 l_1 = d_2 l_2 = d \ l \tag{5.34}$$

where l_1 and l_1 are the distance between ±4 diffraction orders, as shown in in Figures 5.14(b) and 5.14(c), d_1 and d_2 are the corresponding diameters of the wire, and l and d are the distances between ±4 diffraction orders and the diameter of the wire being monitored, respectively.

Example 5.7.2
Determine the diameter of a moving wire in a wiresaw equipment using the diffraction equipment in Figure 5.14. A He–Ne laser was employed with the length between the wire and the screen being $L = 0.6$ m. The separations between ±4 diffraction orders of the patterns are measured are $l_1 = 17.4$ mm, and $l_2 = 20.8$ mm, respectively.

A wire of unknown diameter is found to have a separations between ± 4 diffraction orders of $l = 19.2$ mm. Determine the diameter of the wire being monitored.

Solution:

The He–Ne laser has a wavelength of $\lambda = 632.8$ nm. Applying Equation (5.34), we find

$$d = \frac{l_1 d_1}{l} = \frac{(17.4)(175)}{19.2} = 158.6 \text{ μm} \quad \text{or} \quad d = \frac{l_2 d_2}{l} = \frac{(20.8)(146)}{19.2} = 158.2 \text{ μm.}$$

The discrepancy (less than 0.3%) in the calculation of the diameter of the wire is due to the errors of measurement with a resolution of 0.1 mm and numerical round-off.

Note that it would be incorrect to employ a linear interpolation of the change in wire diameter, as in the following equation

$$\frac{(l_1 - l)}{(l_1 - l_2)} \neq \frac{(175 - d)}{(175 - 146)} = \frac{d_1 - d}{d_1 - d_2} \quad \text{or} \quad d \neq 175 - 29 \left(\frac{l - l_1}{l_2 - l_1} \right).$$

This is due to the fact that l or Δ is inversely proportional to the diameter; therefore, the linear interpolation such as the equation above will not produce the correct results. Instead, the following equation must be used

$$\frac{(l_1 - l)}{(l_1 - l_2)} = \frac{(\Delta_1 - \Delta)}{(\Delta_1 - \Delta_2)} = \frac{(\frac{1}{d_1} - \frac{1}{d})}{(\frac{1}{d_1} - \frac{1}{d_2})}.$$

After some mathematical manipulation, the diameter can be found by the following equation

$$\frac{1}{d} = \frac{1}{d_1} - \left(\frac{\Delta - \Delta_1}{\Delta_2 - \Delta_1} \right) \left(\frac{1}{d_1} - \frac{1}{d_2} \right). \tag{5.35}$$

Example 5.7.3

Use Equation (5.35) to calculate the diameter of the wire, using the data from the previous example.

Solution:

We substitute the numbers from the previous example into Equation (5.35) to obtain

$$\frac{1}{d} = \frac{1}{175} - \left(\frac{19.2/4 - 17.4/4}{20.8/4 - 17.4/4} \right) \left(\frac{1}{175} - \frac{1}{146} \right) \Rightarrow d = 158.3 \text{μm.}$$

Note $\Delta = l/4$. The result is the same as that in the previous example, with a small error from the measurement and numerical round-off.

In addition to the discussions and examples, one can also use two CCD sensors at a fixed distance apart to monitor two non-neighboring diffraction orders to enhance the sensitivity of this technique. In this way, the sensitivity is increased by N times, where

$$N = N_1 - N_2$$

with N_1 and N_2 being the orders selected.

5.7.2 Monitoring the Pitch of the Wire Web Spacing

One of the advantages of the wiresaw machine is the large throughput in producing many wafer slices simultaneously by slicing a long ingot to obtain an array of 200–300 wafers at the same time. This is accomplished by winding a single wire over the surfaces of wire guides, with equally spaced grooves, to form a *wire web* with parallel wire strands (cf. Figures 4.7–4.9). The wafer slices must maintain the consistency of

(i) **Spacing**: uniform thickness across all wafers in one ingot
(ii) **Parallelism**: two highly parallel surfaces on each wafer.

Such consistency must be maintained throughout the slicing operation by ensuring the integrity of the wire web to produce the prescribed spacing and parallelism. Therefore, it is crucial to monitor the wire web and the hundreds of wire strands of the wire web.

Figure 5.15 illustrates a wire strand sitting in a groove on the surface of a wire guide. The wire guide is a cylindrical drum with surface coated by a layer of polyurethane (or similar) material that have evenly spaced grooves, such as the one in Figure 5.15. During the machining process of a slurry wiresaw, the abrasive grits in the slurry will cut through the crystal ingot, as well as the grooves on the surface of wire guides. The wear of the groove illustrated in Figure 5.15(b) is an even wear as the polyurethane material is cut downwards by the wire and abrasive grits in slurry. This is a desirable wear. However, wear of grooves sometime can be uneven, such as the pattern shown in Figure 5.15(c) in which the wear is slanted at an angle, θ, as the downward wear skews to one direction of the supposedly vertical groove. Such uneven wear, as well as patterns of wear other than even wear, are undesirable and will cause the wire web to be distorted to lose the consistency of spacing and parallelism mentioned above. Therefore, it is important to monitor

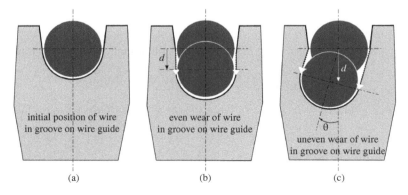

| (a) | (b) | (c) |

Figure 5.15 Illustration of a wire in a groove on the surface of wire guide (only one strand of wire along a wire web is shown here) with different wear of the groove on the surface of wire guide during wiresaw operation: (a) starting initial position of a new groove on surface of a wire guide, (b) desirable and even wear of groove during a wiresaw operation, and (c) undesirable and uneven wear of groove during a wiresaw operation.

Figure 5.16 Moiré fringes created by two superimposed gratings for the detection of pitch and/or orientation mismatches: (a) mismatch in pitch, (b) mismatch in orientation, and (c) mismatch in both pitch and orientation.

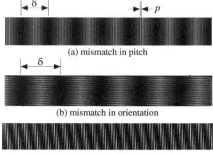

(a) mismatch in pitch

(b) mismatch in orientation

(b) mismatch in both pitch and orientation

the integrity of the wire web to prevent such distortion from happening in slurry wiresaw operations.

An optical system using the moiré technique can be employed to monitor the spacing and parallelism of wire web in real-time by using moiré fringes generated by optical moiré interferometry. Moiré fringes are created when two coarse gratings (the so-called Ronchi gratings) are superimposed with either a difference in pitch, orientation, or both. The schematic and experimental setup for the optical moiré interferometry is presented in Section 10.4.1. Readers may want to first review Section 10.4.1 and related materials to become familiar with the methodology.

Figure 5.16 illustrates different patterns of moiré fringes generated by mismatch in pitch, in orientation and in both pitch and orientation. For example, the uneven wear pattern in Figure 5.15(c) will cause a mismatch in pitch of the wire web at the location of the the uneven wear. This technique is particularly sensitive to mismatch in orientation, as discussed in Figure 10.2 in Section 10.4.1. Indeed, it can be shown [Chiang (1978)] that the angle of mis-orientation θ is magnified by

$$|\phi| \cong \frac{\delta}{p}|\theta|$$

where p is the pitch of the gratings, δ is the spacing between two adjacent pale or dark moiré fringes, and ϕ is the resulting moiré fringe orientation angle measured from the grating lines.

For a given θ and p, both δ and ϕ are determined, but δ is usually several orders of magnitude larger than p. Thus, this phenomenon can be exploited to check the spacing and parallelism of the wire web. Figure 5.17 shows a computer generated moiré pattern with one of the grating having a small sinusoidal variation. It is seen that a slight variation in wire straightness will be greatly enlarged by the moiré phenomenon[9]. Such moiré technique can be employed to detect the linear and/or rotational distortion of wire web using a standard grating of prescribed spacing.

---

9 Note the moiré pattern will be affected by the resolution of printing. For example, viewing it on screen will give different pattern from printed copy. Depending on the resolution of printer, the moiré fringes may not correspond to the real pattern. A different file format, such as a PDF file, may also have affected the resolution. Nevertheless, the illustration of very different pattern can always be appreciated regardless of printing resolution.

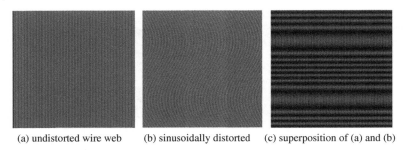

(a) undistorted wire web   (b) sinusoidally distorted   (c) superposition of (a) and (b)

Figure 5.17 Moiré fringes generated by distortion of wire alignment in a wire web as simulated by computer with gratings: (a) undistorted wire web, (b) a wire web which is sinusoidally distorted, and (c) moiré fringes created by superimposing the wire webs in (a) and (b).

5.7.3 Mixing Ratio of Slurry Consisting of Abrasive Grits and Carrier Fluid

The management of slurry, consisting of abrasive grits and carrier fluid, is important in successful operations of slurry wiresaws. The mixing ratios of abrasive grits and carrier fluid, properties of abrasive grits, properties of carrier fluid (water-soluble or oil-based; cf. Section 3.8.2), recycling of the used slurry, and reclaiming the abrasive grits in clurry (cf. Section 3.8.3) are important topics of consideration for successful industrial operations. Readers may want to refer to Section 3.8 for details.

Abrasive grits and carrier fluid are typically mixed with a prescribed ratio between the mass or weight of the abrasive grits (in kilograms) and the volume of the carrier (in liters). The slurry mixing ratio was defined in Equation (3.1), typically in the units of kilograms of abrasive grits (such as silicon carbide) versus liters of carrier fluid (such as glycol). For example with one kilogram of silicon carbide and one liter of glycol, the slurry mixing ratio will be $r_s = 1.0$ kg L^{-1}. The slurry mixing ratio can affect the outcomes of slicing, and are often regarded as an intellectual property of a company when an optimal rate is obtained through the experience of many slicing operations with specific wiresaw equipment and crystal ingots.

5.8 Summary

In this chapter, we presented the rolling-indenting model as a mechanism of slicing and manufacturing process in slurry wiresaw equipment. The vibration modeling and analysis was presented with a historical review of the research and development in this new vibration problem of a moving wire. Practical analysis and evaluation of the damping factor of slurry was presented to guide the analysis of damping in analyses pertaining to vibration of wire and slurry. After that, the treatment of elasto-hydrodynamic (EHD) process modeling was presented that includes the hydrodynamic effect of an elastic moving media in slurry with lubricating effect, as well as the floating machining modeling under the rolling-indenting mechanism. Thermal effect with temperature fluctuation has a profound effect on the quality of wafer surfaces; hence, the thermal management of the equipment is an important subject for practitioners. The real-time measurement of wire wear during wafering using the moiré technique was introduced. Finally, the management of wire, wire web, and slurry was discussed. The measurement of parallelism and constant spacing of wire web with the arrangement of parallel lines is most suitable for the moiré technique. The moiré technique can be used to detect irregular alignment and pattern of wire web to ascertain the integrity of the wire web.

References

Archibald FR and Emslie AG 1958 The vibration of a string having a uniform motion along its length. *ASME Journal of Applied Mechanics* **25**, 347–348.

Bhagavat M, Prasad V and Kao I 2000 Elasto-hydrodynamic interaction in the free abrasive wafer slicing using a wiresaw: Modeling and finite element analysis. *Journal of Tribology* **122**(2), 394–404.

Bhagavat M, Yang F and Kao I 1998 Elasto-plastic finite element analysis of indentations in free abrasive machining *Proceedings of the Manufacturing Engineering Division, IMECE'98*, pp. 819–824. ASME Press.

Bhagavat S and Kao I 2006 Ultra-low load multiple indentation response of materials: In purview of wiresaw slicing and other free abrasive machining (FAM) processes. *International Journal of Machine Tools and Manufacture* **46**(5), 531–541.

Bhagavat S, Liberato J and Kao I 2005 Effects of mixed abrasive slurries on free abrasive machining processes *Proceedings of the 2005 ASPE Conference* ASPE.

Bhagavat S, Liberato J, Chung C and Kao I 2010 Effects of mixed abrasive grits in slurries on free abrasive machining (FAM) processes. **50**, 843–847.

Butkovskiy AG 1983 *Structural Theory of Distributed Systems*. Ellis Horwood Limited, West Sussex, England.

Cameron A 1966 *The Principles of Lubrication*. Wiley, New York.

Chen L 2005 Analysis and control of transverse vibrations of axially moving strings. *Applied Mechanics Reviews* **58**(1-6), 91–115.

Chiang FP 1978 *Manual on Experimental Stress Analysis* Society for Experimental Mechanics chapter 3.

Chung C and Kao I 2008 Damped vibration response at different speeds of wire in slurry wiresaw manufacturing operations *Proceedings of International Manufacturing Science and Engineering Conference (MSEC 2008)* number ASME Paper number MSEC2008-72213 ASME.

Chung C and Kao I 2011 Modeling of axially moving wire with damping: Eigenfunctions, orthogonality and applications in slurry wiresaws. *Journal of Sound and Vibration* **300**, 2947–2963.

Chung C and Kao I 2012 Green's function and forced vibration response of damped axially moving wire. *Journal of Vibration and Control* **18**(12), 1798–1808.

Gohar R 1988 *Elastohydrodynamics* Ellis Horwood Series in Mechanical Engineering.

Gupta P, Kulkarni MS, Zavattari C and R.Vandamme R 2011 Wiresaw ingot slicing systems and method with ingot preheating, web preheating, slurry temperature control and/or slurry flow rate control. Technical report, MEMC Electronic Materials, US Patent No. 7,878,883.

Hamrock B 1994 *Fundamentals of Fluid Film Lubrication*. McGraw-Hill Inc.

Huang FY and Mote CD 1995 On the translating damping caused by a thin viscous fluid layer between a translating string and a translating rigid surface. *Journal of Sound and Vibration* **181**(2), 251–260.

Ishikawa KI and Suwabe H 1987 A basic study on vibratory multi-wiresawing. *Bulletin of the Japan Society of Precision Engineering* **21**(4), 293–295.

Kao I 2002 The technology of modern wiresaw in silicon wafer slicing for solar cells *Invited talk at the 12th Workshop in Crystalline Silicon Solar Cell Materials and Processes*, Brechenridge, Colorado.

Kao I 2004 Technology and research of slurry wiresaw manufacturing systems in wafer slicing with free abrasive machining. *the International Journal of Advanced Manufacturing Systems* **7**(2), 7–20.

Kao I and Bhagavat S 2007 *Single-crystalline silicon wafer production using wire saw for wafer slicing* Transworld Research Network Editors J. Yan and J. Patten, Kerala, India chapter 7, pp. 243–270.

Kao I, Bhagavat M and Prasad V 1998a Integrated modeling of wiresaw in wafer slicing *NSF Design and Manufacturing Grantees Conference*, pp. 425–426, Monterey, Mexico.

Kao I, Chung C and Rodriguez R 2010 *Modern Wafer Manufacturing and Slicing of Crystalline Ingots to Wafers Using Modern Wiresaw* Springer Handbook of Crystal Growth; *ed.* G. Dhanaraj M. Dudley, K. Byrappa, and V. Prasad Springer-Verlag chapter 52.

Kao I, Prasad V, Chiang FP, Bhagavat M, Wei S, Chandra M, Costantini M, Leyvraz P, Talbott J and Gupta K 1998b Modeling and experiments on wiresaw for large silicon wafer manufacturing *the 8th Int. Symp. on Silicon Mat. Sci. and Tech.*, p. p.320, San Diego.

Kao I, Prasad V, Li J and Bhagavat M 1997a Wafer slicing and wire saw manufacturing technology *NSF Grantees Conference*, pp. 239–240, Seattle, Washington.

Kao I, Prasad V, Li J, Bhagavat M, Wei S, Talbott J and Gupta K 1997b Modern wiresaw technology for large crystals *Proceedings of ACCGE/east-97*, Atlantic City, NJ.

Kao I, Wei S and Chiang FP 1998c Vibration of wiresaw manufacturing processes and wafer surface measurement *NSF Design and Manufacturing Grantees Conference*, pp. 427–428, Monterey, Mexico.

Li J, Kao I and Prasad V 1997 Modeling stresses of contacts in wiresaw slicing of polycrystalline and crystalline ingots: Application to silicon wafer production *Proceedings of ASME IMECE '97*, pp. 439–446. ASME Press, Dallas, Texas.

Li J, Kao I and Prasad V 1998 Modeling stresses of contacts in wiresaw slicing of polycrystalline and crystalline ingots: Application to silicon wafer production. *Journal of Electronic Packaging* **120**(2), 123–128.

Ma F, Morzfeld M and Imam A 2010 The decoupling of damped linear systems in free or forced vibration. *Journal of Sound and Vibration* **329**(15), 3182–3202.

Ma JTS 1965 An investigation of self acting foil bearings. *Transactions of the ASME, Journal of Basic Engineering* **87**, 837–846.

Malookani RA and van Horssen WT 2015 On resonances and the applicability of Galerkin's truncation method for an axially moving string with time-varying velocity. *Journal of Sound and Vibration* **344**(0), 1–17.

Meirovitch L 1997 *Principles and Techniques of Vibrations* 1st edn. Prentice Hall.

Mizuno M, Iyama T, Kikuchi S and Zhang B 2006 Development of a device for measuring the transverse motion of a saw-wire. *Journal of Manufacturing Science and Engineering, Transactions of the ASME* **128**, 826–834.

Oh K and Huebner K 1973 Solution of the elastohydrodynamic finite journal bearing problem. *Transactions of the ASME. Journal of Lubrication Technology* pp. 342–352.

Oz HR and Pakdemirli M 1999 Vibrations of an axially moving beam with time-dependent velocity. *Journal of Sound and Vibration* **227**(2), 239–257.

Pakdemirli M and Ulsoy A 1997 Stability analysis of an axially accelerating string. *Journal of Sound and Vibration* **203**(5), 815–832.

Pakdemirli M, Ulsoy AG and Ceranoglu A 1994 Transverse vibration of an axially accelerating string. *Journal of Sound and Vibration* **169**(2), 179–196.

Papanastasiou TC 1994 *Applied Fluid Mechanics*. Prentice-Hall.

Ponomareva S and van Horssen W 2009 On the transversal vibrations of an axially moving continuum with a time-varying velocity: Transient from string to beam behavior. *Journal of Sound and Vibration* **325**(4-5), 959–973.

Roach GF 1982 *Green's Functions* 2nd edn. Cambridge University Press, Cambridge, United Kingdom.

Rohde S and Oh K 1975 A unified treatment of thick and thin film elastohydrodynamic problems by using higher order element methods. *Proc. R. Soc. Lond. A* **343**, 315–331.

Rohde S and Oh K 1977a Numerical solution of point contact problem using the finite element method. *International Journal for Numerical Methods in Engineering* **11**, 1507–1518.

Rohde S and Oh K 1977b A theoretical analysis of a compliant shell air bearing. *Transactions of the ASME. Journal of Lubrication Technology.* **24**, 75–81.

Sahoo R, Prasad V, Kao I, Talbott J and Gupta K 1998 Towards an integrated approach for analysis and design of wafer slicing by a wire saw. *Journal of Electronic Packaging* **120**(1), 35–40.

Sherman FS 1990 *Viscous Flow*. McGraw-Hill.

Swope RD and Ames WF 1963 Vibration of a moving threadline. *Journal of Franklin Institute* **275**, 36–55.

Szeri A 1980 *Tribology: Friction, Lubrication, and Wear* 1st edn. Hemisphere Publishing Co., ISBN 0070626634.

Tipei N 1978 Flow characteristics and pressure head build-up at the inlet of narrow passages. *Transactions of the ASME. Journal of Lubrication Technology* **100**, 47–55.

Tipei N 1982 Flow and pressure head at the inlet of narrow passages, without upstream free surface. *Transactions of the ASME. Journal of Lubrication Technology* **104**(2), 196–202.

Toyama K, Kiuchi E and Hayakawa K 1992 Wire saw and slicing method using the same. Technical report, Shin-Etsu Handotai Company, Ltd., US Patent No. 5,269,285.

van Horssen WT 2003a On the influence of lateral vibrations of supports for an axially moving string. *Journal of Sound and Vibration* **268**, 323–330.

van Horssen WT 2003b On the influence of lateral vibrations of supports for an axially moving string. *Journal of Sound and Vibration* **268**, 323–330.

van Horssen WT and Ponomareva SV 2005a On the construction of the solution of an equation describing an axially moving string. *Journal of Sound and Vibration* **287**, 359–366.

van Horssen WT and Ponomareva SV 2005b On the construction of the solution of an equation describing an axially moving string. *Journal of Sound and Vibration* **287**, 359–366.

Wei S and Kao I 1998 Analysis of stiffness control and vibration of wire in wiresaw manufacturing process *Proceedings of the Manufacturing Engineering Division, IMECE 98*, pp. 813–818. ASME Press.

Wei S and Kao I 2000a Free vibration analysis for thin wire of modern wiresaw between sliced wafers in wafer manufacturing processes *the Proc. of IMECE'00: EEP-Vol 28: Packaging of electronic and photonic devices*, pp. 213–219. ASME Press, Orlando, Florida.

Wei S and Kao I 2000b Vibration analysis of wire and frequency response in the modern wiresaw manufacturing process. *Journal of Sound and Vibration* **231**(5), 1383–1395.

Wei S and Kao I 2004 Stiffness analysis in wiresaw manufacturing systems for applications in wafer slicing. *the Int. Journal of Advanced Manufacturing Systems* **7**(2), 57–64.

Wickert JA 1994 Response solutions for the vibration of a traveling string on an elastic foundation. *Journal of Vibration and Acoustics* **116**, 137–139.

Wickert JA and Mote CD 1988 Current research on the vibration and stability of axially moving materials. *Shock Vibration Digest* **20**, 3–13.

Wickert JA and Mote CD 1990 Classical vibration analysis of axially moving continua. *ASME Journal of Applied Mechanics* **57**, 738–744.

Wickert JA and Mote CD 1991 Response and discretization methods for axially moving materials. *ASME Applied Mechanics Review* **44**, S279–S284.

Wu S, Wei S, Kao I and Chiang FP 1997 Wafer surface measurements using shadow moiré with Talbot effect *Proceedings of ASME IMECE'97*, pp. 369–376. ASME Press, Dallas, Texas.

Yang F and Kao I 1999 Free abrasive machining in slicing brittle materials with wiresaw. Technical Report TR99-03, SUNY at Stony Brook, Department of Mechanical Engineering, Stony Brook, NY 11794-2300.

Yang F and Kao I 2001 Free abrasive machining in slicing brittle materials with wiresaw. *Journal of Electronic Packaging* **123**, 254–259.

Ying S and Tan CA 1996 Active vibration control of the axially moving string using space feedforward and feedback controllers. *Journal of Vibration and Acoustics* **118**(3), 306–312.

Zavattari C, Severico F, Bhagavat S, Vercelloni G and Vandamme RR 2013 Methods for controlling surface profiles of wafers sliced in a wire saw patent application. Technical report, MEMC Electronic Materials, US Patent No. 20130139800.

Zhu L and Kao I 2005a Computational model for the steady-state elasto-hydrodynamic interaction in wafer slicing process using wiresaw. *the Int. Journal of Manufacturing Technology and Management* **7**(5/6), 407–429.

Zhu L and Kao I 2005b Galerkin-based modal analysis on the vibration of wire-slurry system in wafer slicing using wiresaw. *Journal of Sound and Vibration* **283**(3-5), 589–620.

Zhu L, Bhagavat M and Kao I 2000 Analysis of the interaction between thin-film fluid hydrodynamics and wire vibration in wafer manufacturing using wiresaw *the Proc. of IMECE'00: EEP-Vol 28* packaging of electronic and photonic devices, pp. 233–241. ASME Press, Orlando, Florida.

6

Diamond-Impregnated Wire Saws and the Sawing Process

Wafer manufacturing encompasses a plethora of processes and machine tools to produce prime wafers for microelectronic fabrication, as presented in Chapter 2 of this book. Included in wafer manufacturing processes are crystal growth, wafer forming, wafer polishing and wafer preparation. Modern slurry wiresaws and technology are introduced in Chapter 4, with a comparison to the inner-diameter (ID) saws. Today, wiresaws have become the most common machine tool for slicing ingots of various materials including silicon, sapphire, silicon carbide, and other crystalline materials into wafers for various applications, including semiconductor, photovoltaic, and mobile phones. It was only very recently that the diamond wire saws became a competitive technology to slurry wiresaws as a result of improvements in materials technology and of bonding diamond abrasive grits being placed onto the circumferential surface of a wire with very small diameter. In this chapter, *diamond-impregnated wire saws* (commonly referred to as simply the *diamond wire saws*) will be introduced and presented.

6.1 Introduction

Diamond wire saws have been used in stone slicing since 1978 [Huang et al. (2009)]. In this application of stone quarrying, diamond beads with abrasive grits on them are mounted on a steel wire with spacers, as illustrated in Figure 6.1. The beads are impregnated with diamond grits as abrasives. The motion of the steel wire, which carries the diamond beads and spacers between them, provides abrasion on the surface of the stone to remove stone chips from the surface to slice through the stone.

In wafer slicing using wire saws, the diameter of the wire should be as small as possible in order to reduce the kerf loss. Under such consideration, a steel wire coated with a layer of diamond abrasive grits on the circumferential surface of the wire is an appealing option as a tool for wafer slicing. Figure 6.2 shows SEM images of a diamond wire with a diameter of 150 μm used in diamond wire saws for slicing silicon wafers. Diamond-impregnated wire saws have attracted attention after the success of slurry wiresaws in the last decades. However, the lack of a proven

Wafer Manufacturing: Shaping of Single Crystal Silicon Wafers,
First Edition. Imin Kao and Chunhui Chung.
© 2021 John Wiley & Sons Ltd. Published 2021 by John Wiley & Sons Ltd.

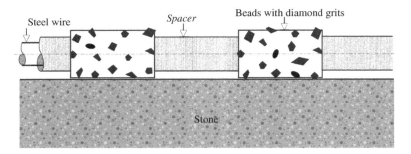

Figure 6.1 A steel wire with beads impregnated by diamond abrasive grits, mounted on a steel wire with spacers for stone quarrying.

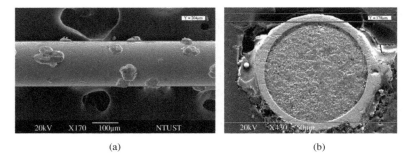

Figure 6.2 SEM images of a diamond wire showing (a) a view of the wire's circumferential surface, and (b) a cross-sectional view with the outer nickel-electroplated layer. Source: Chunhui Chung.

bonding technology for wires had prevented a wide adoption of diamond wire saws for wafer slicing until recent significant improvements of wire bonding technology.

Today, there are two prevailing wire bonding technologies for diamond wires. They are:

1. Resin bonding
2. Electroplating.

Resin bonded diamond wires are low-cost. Furthermore, kilometer-long wires can be easily and consistently produced. However, resin bonding results in weak bonding strength between the steel wire and its diamond coating, deterring its widespread utilization in wafer slicing process. Because the bond between the steel wire and the diamond coating is not strong enough, diamond abrasive grits can be easily stripped from the surface of the wire, reducing the ability of the diamond wire to slice through ingot. The resin bonded diamond wires, however, have higher breaking twist strength and breaking bending strength, as in Table 6.1.

On the other hand, nickel-electroplated diamond wires have been gaining widespread usage in wafering processes for slicing ingots to wafers. Table 6.1 compares diamond wires made with these two different bonding technologies. The electroplated diamond wire has lower breaking strength due to the larger thickness and higher shear modulus of the coating on the electroplated

Table 6.1 Comparison of resinoid and electroplated diamond wires.

Bonding type	Resin bond	Electroplating
Cost	Low	High
Bonding strength	Low	High
Heat resistance	Low	High
Breaking twist strength	High	Low
Breaking bending strength	High	Low

layer [Enomoto et al. (1999)]. In order to reduce production costs, Chiba et al. (2003) proposed a high-speed manufacturing method to increase the production efficiency of the electroplated diamond wire in 2003. Taniguchi et al. (2010) patented an electro-deposited wire tool, in which a very thin layer of coating is formed on the super abrasive grains to increase bonding ability. This technology may also reduce the thickness of the coating layer and increase the breaking strengths, although these were not emphasized in the patent. Figure 6.2 shows SEM images of nickel-electroplated diamond wire. In this figure, the circumferential surface of a diamond wire is shown with visible diamond abrasive grits and the cross-sectional area of the wire with the outer nickel-electroplated layer.

Reports of successful slicing using a diamond wire saw appeared after 2000. Diamond wire saws were used to cut very hard and brittle materials such as sapphire, silicon carbide (SiC), or various semiconductor compound materials. Clark et al. (2003a,b) reported their research results on the performance and monitoring of diamond wire sawing processes of wood and ceramic foam. Hardin et al. (2004) and Huang et al. (2015) presented their experimental results for using a diamond wire saw to slice SiC wafers. Other researchers further reported the wire sawing of sapphire, such as [Ide (2007); Meng et al. (2006); Teomete (2011a,b)]. While bonded abrasive machining (BAM) using diamond wire saws can reduce the processing time to one quarter of the time required for slurry wiresawing, which uses free abrasive machining (FAM), it was not seen as suitable for PV wafering because a very long and uniform wire is required to cut hundreds or thousands of wafers in one run. It was challenging to produce a very long diamond coated wire at the beginning of the 21st century [Pauli et al. (2005)]. Studies on etching and texturing of a diamond wire sliced PV wafer surface appeared in 2009 [Bidiville et al. (2009); Chen et al. (2014); Lippold et al. (2014)], which suggested the success of diamond wire saws in PV wafer slicing.

6.2 Manufacturing Processes of Diamond-impregnated Wires

In the following sections, the two manufacturing processes to produce diamond wires, using the techniques of resin bonding, or electroplating, as mentioned in Section 6.1, will be discussed in more detail.

Table 6.2 Manufacturing conditions of a resinoid diamond wire.

Product diameter	0.26–0.28 mm
Core wire	0.2 mm diameter piano wire
Bonding material	Liquid phenolic resin (resol type)
Abrasive grain	30–40 µm diamond

6.2.1 Resinoid Wires

The fundamental manufacturing processes to produce resinoid diamond wire are similar to those to produce composite materials, namely, with polymers reinforced by small hard particles. Enomoto et al. (1999) reported a manufacturing process of resinoid diamond wire. The manufacturing conditions are listed in Table 6.2 [Enomoto et al. (1999)]. The core wire is a steel piano wire with a diameter of 0.2 mm. The bonding material is a liquid phenolic resin. Abrasive grain are diamond abrasive grits of size from 30 to 40 µm. The concentration of the abrasive grains in resin is 50% by weight. Ide (2007) provided a wider range of the manufacturing conditions of the resinoid diamond wire. Depending on different slicing operations, the wire with different conditions would be considered. The manufacturing procedures provided by Enomoto et al. (1999) are enumerated in the following:

1. Diamond grains are first stirred to mix in liquid resin.
2. The surface of the core wire is sanded with #600 sandpaper to improve adhesion between the wire and the resin.
3. The wire surface is cleaned with acetone.
4. The core wire is coated with the bonding resin, which now includes diamond grains immersed.
5. The wire is dried at 70 °C for 60 min and cured at 150 °C for 10 min.

Figure 6.3 illustrates the manufacturing apparatus of resinoid diamond wire. The apparatus is similar to drawing manufacturing equipment, with a reservoir of liquid

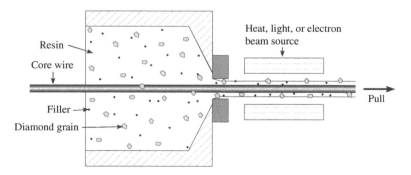

Figure 6.3 Illustration of the manufacturing apparatus of resinoid diamond wire.

resin containing the mixture of diamond grains and filler particles. As the core wire is pulled, it carries with it the resin mixture in the reservoir through a gradually converging passage way to cause the diamond grains and filler particles to coat onto the surface of the wire along with resin. As described earlier, the pulled wire coated with resin mixture is dried and cured for the final production of the resinoid diamond wire.

In order the improve the strength and performance of the resinoid diamond wire, Enomoto et al. (1999) fabricated diamond wires using the aforementioned procedures, except by adding different micro-powders as filler in the resin at the first step of stirring and mixing. The different micro-powders included: copper powder (0.94 μm), silicon carbide powder (1 μm), aluminum oxide powder (1.2 μm), and diamond powder (0–1 μm). The experimental results show that the resinoid diamond wire reinforced by 60wt.% of copper powder in the bonding resin is superior to the resinoid diamond wire in breaking twist strength, resistance to heat, slicing performance, and tool wear. The addition of the copper powder restrains the carbonization of the core wire and prevents the generation of blue brittleness[1]. This improves the breaking twist strength. The high heat conductivity of copper increases the resistance of heat of the resinoid diamond wire. The holding strength of abrasive grains was supposed to be increased by adding copper powder, which improves the tool wear resistance as well as the slicing performance.

Based on the same concept to improve the performance of resinoid diamond wire by adding fine powders, Ueoka et al. (2000) and Shimazaki et al. (2002) patented a resinoid diamond wire with filler of different powders at different concentrations and grit sizes. A superior slicing performance was obtained.

Shimazaki et al. (2002) proposed a method to increase the manufacturing speed of the resinoid diamond wire by using a light curing resin or an electron beam curing resin as the bonding material. The diamond wire bonded by the light curing resin can be continuously fabricated at a speed of a few hundred meters per minute to a few kilometers per minute since the light curing resin can be cured within a few seconds. However, the curing may not be fully completed due to transmission, absorption or reflection of light at or near the particles of the fillers. The solution proposed in the same patent is to use an electron beam curing resin. In addition to better curing completion, electron beam curing resin can be cured within an even shorter time than the light curing resin. The additional curing units using heat, light, or electron beam source, with corresponding resins, are illustrated in Figure 6.3. A bonding method of using a primer layer between the core wire and the light curing resin layer was also proposed in the same patent [Shimazaki et al. (2002)] to securely adhere the light curing resin layer to the core wire. A photopolymerization accelerator or a coupling agent such as silane can be used as the primer layer.

The researchers at the Noritake Super Abrasive Cooperation published reports of slicing sapphire and silicon ingots using resinoid diamond wire [Ide (2007);

---

1 Blue brittleness is a specific brittleness exhibited by some steels after being heated to temperatures within the range 300–650 °F (149–343 °C), particularly if the steel is worked at such elevated temperatures.

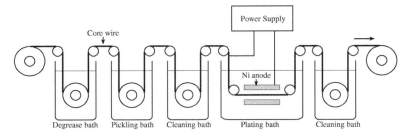

Figure 6.4 Illustration of the manufacturing apparatus of electroplated diamond wire.

Watanabe et al. (2010)]. Ide (2007) claimed that the resinoid diamond wire has lower machining efficiency and lower tool life compared to the electroplated diamond wire. However, the sliced wafer surface and the flexibility (such as the strength of twisting, defined by the number of twists) are better. Watanabe et al. (2010) studied the slicing of PV polycrystalline silicon wafer using resinoid diamond wire, with a size of 50×50 mm and a feed rate of 0.8 mm min^{-1}.

6.2.2 Electroplated Wires

Nickel-electroplating is the prevailing method in producing diamond wires because of the high wear resistance of nickel plating layer. During the electroplating process, the core steel wire has to pass through several chemical baths to produce a well bonded diamond wire. An illustration of the production of an electroplated diamond wire is illustrated in Figure 6.4. The figure illustrates typical procedures, including

1. Degreasing: The core wire first passes through the first bath to remove grease on the wire, which may be coated during the traditional wire drawing manufacturing process to produce the cold-drawn steel wire.
2. Pickling: The core wire is then immersed in the second bath to remove any residue.
3. Cleaning: The next bath cleans the core wire ready for electroplating.
4. Electroplating: The core wire passes through the electroplating bath, with power supply configuring Ni anode and the core wire cathode for electroplating. The diamond abrasive grains are mixed into the electroplating bath. The plating bath normally is composed of nickel sulfamate, nickel chloride, and boric acid [Chiba et al. (2003); Ge et al. (2009); Taniguchi et al. (2010)]. The current density of the power supply will affect the thickness of deposited layer.
5. Second cleaning: The second cleaning bath performs the final cleansing of the electroplated diamond wire.

The five steps of the procedures to produce electroplated diamond wires illustrated here can embody different degree of details and variations. However, the procedures are largely the same for producing electroplated diamond wires employed in the diamond wire saws.

Electroplating is a process commonly used in the manufacturing industry for increasing lifetime of tools or preventing corrosion. For the purposes of understanding the electroplating process, we present a simple example of coating a metal with another metal. In this example, a suitable electroplating setup consists of an electrode to be coated, a source of coating metal, a current source with a power supply, and an electrolyte solution. In the electroplating process, the electrolyte solution, rich in metal cations (positive ions), utilizes electric current to reduce the metal cations onto the electrode surface. The thickness, uniformity, conformity, and overall quality of the metal coating are highly dependent on both electrical parameters (current, voltage, cell resistance, etc.), as well as chemical parameters (concentration of electrolyte, metal source, etc.).

For example, when plating copper onto a conductive substrate, one can use a copper metal source and a copper sulfate electrolyte. The copper sulfate electrolyte, rich in both copper cations (positive ions) and sulfate anions (negative ions), provides the cations that will ultimately reduce onto and coat the substrate. If the conductive substrate is negatively biased while submerged in the electrolyte, it will reduce the positive ions (cations) in solution. Likewise, when the copper metal source is positively biased while submerged in the electrolyte, copper will oxidize (become cations) and associate with the anions in solution, replenishing the source of copper within the electrolyte.

Since diamond is a dielectric material, the diamond grains will increase the resistance in the electroplating bath, thereby decreasing the plating efficiency and making it difficult to electroplate a solution. Various methods have been proposed to plate diamond efficiently, most of which use nickel in the electrolyte solution or coat diamond abrasives with nickel, in order to increase the conductivity and quality of film. Other conductive metals, such as copper, have been studied; however, they have shown worse bonding properties in general. Although copper has better electric conductivity by itself, the use of copper as conductive metal results in weaker bonding in the electroplating process [Chiba et al. (2003)]. The use of nickel in diamond electroplating processes is widespread, although recipes for electroplating diamond are proprietary and closely guarded by industry.

6.2.3 Machines and Operations of Diamond Wire Saws

Machining using diamond wire saws, utilizing diamond-impregnated (or diamond-coated) wires, belongs to abrasive machining, as discussed in Section 3.4. Specifically, the abrasive grits are bonded onto the machine tool (in this case, the wire), resulting in a bonded abrasive machining, or BAM, as presented in Section 3.4.1.

There are different designs of wire saw machine tools using diamond-impregnated wires. As an example, a single-pass, reciprocating saw using a wire impregnated with industrial diamond grits was shown in Figure 2.13. This wire saw is designed more for the laboratory instead of industrial applications.

Industrial diamond wire saws are similar to slurry wiresaws for wafer slicing and can be adapted from the design of slurry wiresaws. The wire saw equipment manufacturers sell diamond wire saw machine tools under different labels or models. However, some of the components and processes in operation and control must be re-designed due to the nature of the diamond-impregnated wire. The grains coated on the diamond-impregnated wire are abrasive grits, which will cause more wear and tear on the machine components that are in direct contact with the diamond wire when compared with slurry wiresaws. For example, the pulleys, used to guide the diamond wire from one bobbin to the machining area and to another bobbin, will be subjected to more direct wear because of the bonded diamond abrasive grits (BAM, cf. Section 3.4.1) on the surface of wire, and will require special consideration and design. Furthermore, the wear and tear of pulleys must be monitored regularly.

Like the slurry wiresaws, the diamond wire saws operate by moving the wire back and forth in order to lengthen the time of operating the wire in slicing operation. However, the frequency of the back-and-forth movement of the wire in a diamond wire saw is higher than that in a slurry wiresaw. In addition, the ratio of wire traveling distance in backward motion to that in forward motion (also called the "B/F ratio") in one back-and-forth movement of the diamond wire is also higher than that in slurry wiresaw. For example, the wire of a diamond wire saw may move 300 m forward and 290 m backward in one back-and-forth movement of the wire, with a high ratio of 0.967[2]. This back-and-forth movement of wire continues until the end of the slicing process. Depending on the tool wear, the spent wire might be used for the next run or discarded after each slicing operation. The high B/F ratio of the diamond wire is due to its longer tool life. Therefore, the diamond wire can sustain longer traveling distance to remove more materials.

The diamond grains typically can work until they are dull, especially with the Ni-electroplated diamond wires. The bonding strength of the Ni-electroplated diamond wire is usually stronger than that of the resinoid diamond wire, which means that less diamond grains come off during slicing. The abrasive grains which came off the diamond-coated wire would penetrate not only to the workpiece but also the wire if not flushed out. In addition, the bonding layer of the bonded-abrasive diamond wire protects the core wire during slicing process. The Ni-electroplated layer outperforms the resinoid layer in the wear resistance [Enomoto et al. (1999)]. The total wire length for one operation is in the range of 2–5 km. These slicing parameters are dependent upon different industry applications, in which the ingot material, ingot size, number of wafers, and operation efficiency must be considered. The back-and-forth tool marks can be observed on the wafers sliced by diamond wire saws, just like the tool marks of wafers sliced by slurry wiresaws.

Diamond wire saws typically operate at a slower wire speed than the slurry wiresaws. This is because the machining mechanism of diamond wire saw is bonded abrasive machining (BAM) process, in which the cutting force is transferred directly to the wire. Operating at too high a wire speed may result in premature rupture of the wire. The wire speed of a diamond wire saw is normally in the range of 8–12 m s^{-1}, although this is still rather high compared to other machining processes.

---

2 Note that this ratio of backward length to forward length is always smaller than one (1.0).

Table 6.3 Comparison between operating properties of diamond wire saws and slurry wiresaws.

Operating properties	Diamond wire saw	Slurry wiresaw
Free abrasive machining process	BAM	FAM
Equipment design	Similar (attention to more wear)	Similar
Abrasive grits	Bonded diamond grits	Free abrasive grits (SiC, B_4C, diamond)
Frequency of back-and-forth motions	Higher	Not as high
Ratio of backward to forward length	Higher	Not as high
Tool marks on wafer surface	Visible	Visible
Wire speed	$8–12\,\text{ms}^{-1}$	$10–20\,\text{ms}^{-1}$
Feed rate	High	Low
Slice material with high hardness	Easier	More difficult

The strategy of the operation of diamond wire saw is using relatively high wire speed with high frequency of the back-and-forth wire motion and low feed rate compared to the wire speed. The feed rate is about the order of $0.1\,\text{mm min}^{-1}$.

Diamond wire saws have been employed to slice materials with high hardness, such as silicon carbide or sapphire. In such situations, low feed rate is required to avoid premature rupture of the diamond wire. The increase in the wire speed and the increase in frequency of the back-and-forth wire motion will increase the number of diamond grains in contact with the workpiece in the same period of time, thereby increasing the efficiency of slicing. The speed of the wire is constrained by the type of material, wire strength, and the machine specification, such as the limitation of driving motor power.

A table with comparison of several operating properties between the diamond wire saws and slurry wiresaws is summarized in Table 6.3.

6.3 Slicing Mechanism of a Diamond Wire Saw

In contrast to slurry wiresaws, which operate under the free abrasive machining (FAM) process with rolling-indenting, as discussed in Sections 3.4 and 3.4.2, the diamond wire saws operate under the bonded abrasive machining (BAM) process, as discussed in Sections 3.4 and 3.4.1. Table 6.4 compares the different types of abrasive machining processes and equipment commonly used in wafer manufacturing. Other than slurry wiresaws, lapping, and some polishing processes that utilize free abrasives in a three-body abrasion and interaction, all other abrasive machining processes are BAM.

Table 6.4 Comparison of various common abrasive machining types in wafer manufacturing.

Abrasive machining type in wafer manufacturing	Bonded abrasive machining (BAM)	Free abrasive machining (FAM)
Diamond wire saw	BAM	–
Slurry wiresaw	–	FAM
Lapping	–	FAM
Polishing	–	FAM
Wafer grinding	BAM	–
ID saw	BAM	–

The metal grinding process can be divided into the following three regimes, depending on the depth of cut of each single abrasive.

- Rubbing: Rubbing corresponds to elastic deformation of metallic workpiece.
- Ploughing: Ploughing corresponds to ridge formation due to elasticity and plastic deformation of the metallic workpiece.
- Micro cutting: Micro cutting results in chip formation from the surface of metal. This is also commonly referred to as the "orthogonal machining model" (cf. Section 3.3).

Of the three regimes of the metal grinding process, only the mechanism of micro cutting removes material from the workpiece. Rubbing does not change the surface because the deformed surface bounces back after the abrasive grits pass through. The ploughing action pushes the material to the side of the cutting groove to form the ridge, which is caused by plastic deformation that modifies the surface morphology.

The process for brittle materials is more complex. Due to the brittle nature of semiconductor and PV materials, chip formation or material removal can be further divided into ductile-regime machining and brittle machining processes (cf. Section 3.3). The critical depth of cut is typically in the order of 0.1 μm, which varies according to the material and orientation of the normal and cutting directions. Once the depth of cut is larger than the critical depth of cut, brittle fracture would take over to be the primary material removal mechanism. Such brittle fracture also leaves fracture marks on the surface. Rubbing and ploughing are not always apparent in the grinding of brittle materials.

On the surfaces of silicon wafers sliced by diamond wire saws, there are many ploughing marks with ridge formation from plastic deformation of brittle material due to the high wire speed and low feed rate. However, various cracks from brittle fracture are also visible on the surface because the diamond abrasives are forcing themselves upon the surface of the workpiece in a BAM machining process when the diamond wire passes through. Figure 6.5(b) presents a typical surface of silicon wafer sliced by a diamond wire saw, as compared with the surface of a wafer sliced by a slurry wiresaw. In fact, the surface sliced by a diamond wire has a lot of microscopic similarities to the surface sliced by an ID saw, as presented in Figure 4.4.

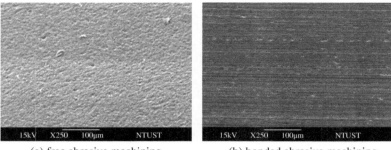

(a) free abrasive machining   (b) bonded abrasive machining

Figure 6.5 SEM images of surfaces of silicon wafers, in the same scale of 100 μm for comparison, produced by (a) a slurry wiresaw using a FAM process, and (b) diamond wire saw using a BAM process. Source: Chunhui Chung.

This is expected because both ID saws and diamond wire saws are BAM processes. Figure 6.5(a) is a very typical surface of a silicon wafer sliced by slurry wiresaws, as presented in Figures 5.2 and 5.13. Figure 6.5(b) is a typical surface of silicon sliced by a diamond wire saw using a BAM process.

During the slicing process using the diamond wire saw, the material directly under the wire along the direction of the feed is under high normal force generated by the wire tension and the bow angle of the wire. Materials are removed as the kerf loss with a high degree of fracture of brittle material. The force applied on the slices of wafer surface is smaller than the normal force in the feed direction of the workpiece (perpendicular to each other). This could be one of the reasons why the as-sliced wafer surface shows only a few fracture marks, although the main material removal mechanism is the brittle fracture [Chung and Nhat (2015)]. Figure 6.6 illustrates the

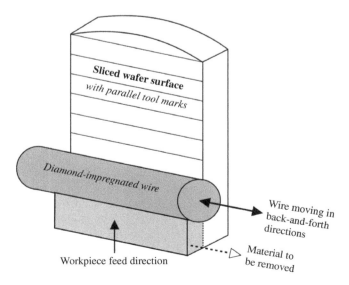

Figure 6.6 Illustration of the diamond wire moving direction and the formation of wafer surface during slicing, with back-and-forth motion.

direction of wire motion and the formation of wafer surface during slicing. While the ingot is fed to the moving diamond wire from bottom up, the material immediately beneath the diamond wire is removed by the diamond wire. Due to the reciprocating motion of the diamond wire, tool marks are left on the wafer surface as the ingot move past the diamond wire, as illustrated in Figure 6.6.

Good models of brittle indentation fracture are important for understanding the brittle material removal in slicing. Nevertheless, the motion of abrasives in the machining process involves both scratching and indenting. Experiments have been conducted to investigate single diamond tip scribing in fly cutting using a turning machine tool to understand the machining mechanisms of brittle materials, especially in ductile-regime machining. Results showed that the critical depth of cut was affected by several parameters including tip shape, moving speed, and surface crystalline orientation. In order to understand the formation of cracks during diamond wire sawing, Wu and Melkote (2012) applied an extended finite element method (XFEM) based model in their study and compared it with experimental results where a single diamond was used to scribe silicon. Their study results showed that increasing the tip radius and/or the included angle increased the critical depth of cut by delaying crack initiation in silicon. Furthermore, lowering the coefficient of friction also yielded a higher critical depth of cut. Their study also demonstrated that the XFEM model, by predicting experimental results, can be used to study crack formation.

Refer to Chapter 3 and Sections 3.3 and 3.6 for more presentations on related subjects and details. Brittle machining is a complex subject and will require further research studies to understand the intertwined nature of interactions with a moving diamond-impregnated wire and materials.

6.4 Properties of Wafers Sliced by Diamond Wire Saws

In the following sections, various properties of wafers sliced by diamond wire saws are presented, and comparison with wafers sliced by slurry wiresaws are provided. The properties include wafer surface, fracture strength, residual stress, PV wafer efficiency, and cost of wafering.

6.4.1 Wafer Surface

The surface morphology of silicon wafers sliced by slurry wiresaws and diamond wire saws is presented in Figure 6.5. The difference in surface morphology arises due to the nature of machining mechanisms; namely, FAM with rolling-indenting and fractures for the slurry wiresaws and BAM with scratching and fractures for the diamond wire saws. Silicon wafers sliced using a slurry wiresaw show randomly distributed pits on the surface, while those sliced by a diamond wire saw show scratch marks with some fracture cracks on the surface. In addition to waviness, silicon wafers sliced by diamond wire saws tend to have wire tool marks that have spatial wave length longer than scratching marks but shorter than the waviness of the wafer [Chen and Chao (2010); Teomete (2011a)]. These tool marks are a result of the

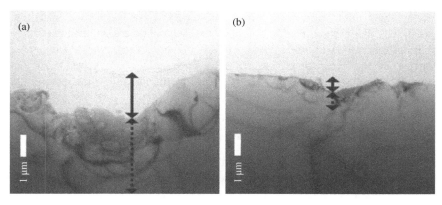

Figure 6.7 Cross-sectional TEM images of surface of a silicon wafer sliced by (a) a slurry wiresaw, and (b) adiamond wire saw. Source: Watanabe et al. (2010).

reciprocating motion of the diamond wire. Such tool marks also appear on the surface of wafers sliced by reciprocating slurry wiresaws [Wu et al. (2014)]. Although investigated under different slicing parameters, such as grain size and geometry, research showed that the surface roughness of silicon wafers sliced by a diamond wiresaw can be better than that of a silicon wafer sliced by a slurry wiresaw [Chen and Chao (2010)], and that the subsurface crack length is shorter [Watanabe et al. (2010); Wu et al. (2012)]. Figure 6.7 compares the crack lengths of wafers sliced by a slurry wiresaw and a diamond wire saw, showing more cracks on the surface and subsurface damage of a wafer sliced by a slurry wiresaw in cross-sectional TEM images. The surface of the wafer sliced by a diamond wire saws appears to have a shorter subsurface radial crack. Wafers sliced by diamond wire also have better performance in taper[3] due to the reciprocating motion and stable diamond abrasive size [Kray et al. (2007); Yang et al. (2013b)].

6.4.2   Wafer Fracture Strength

Recent studies on silicon wafers sliced by slurry wiresaws and diamond wire saws have deepened the understanding of fracture strengths of wafers [Wu et al. (2012); Yang et al. (2013a,b)]. A typical PV silicon wafer has a thickness of 180 μm, with likely fracture during handling and processing of solar cells. Moreover, the fracture strength is correlated with the microcracks in the wafers, which have been shown to be larger at the wafer edge than in the center. According to a study [Wu et al. (2012)], wafers sliced by diamond wire saws tend to have higher crack density but smaller average crack length than wafers sliced by slurry wiresaws. As a result, the experimental results of fracture strength of wafers sliced by diamond wire saws is comparable or better than the strength of wafers sliced by slurry wiresaws. However, wafers sliced by diamond wire saws are prone to leaving deeper scratch tool marks on the

---

3 Taper refers to the thickness difference between both ends of a wafer. This is the linear component of the variation in thickness across a wafer, indicated by the angle between the best fit plane to the front surface and the ideally flat back surface of the wafer. It is also called the "wedge."

surface, which result in different bi-directional strengths; that is, such wafers, when bent perpendicular to the direction of tool marks, will obtain a higher value of fracture strength than the wafer bent parallel to the direction of tool marks [Yang et al. (2013a)]. This may be correlated with the subsurface radial cracks generated along the direction of the scratch tool marks due to the indentation and brittle machining (cf. Section 3.2 and Figure 3.2(b)).

6.4.3   Residual Stress and Stress Relaxation

Yang et al. (2013b) used a full-field near-infrared polariscope to measure residual stress in thin crystalline silicon wafers sliced by diamond wire saws. The average residual stress of as-sliced wafers is 5.61–6.69 MPa, and the local maximum residual stress is 37.5–38.9 MPa. After an etching with a layer of 5 µm material removed, the average residual stress was reduced to 2.86–3.25 MPa while the local maximum residual stress was reduced to 21.9–30 MPa. The fracture strength of the wafers after etching increases significantly due to the reduction of residual stress (stress relaxation) and crack size.

6.4.4   PV Wafer Efficiency

A major consideration for wafers sliced by diamond wire saws is the overly scratched tool marks on wafer surfaces. Since scratch marks typically do not appear on the surface of wafers sliced by slurry wiresaws, slicing PV solar cells using diamond wire saws poses a significantly disadvantage. A recent study by Bidiville et al. (2009) and Lippold et al. (2014) showed that the energy generation efficiency of PV wafers sliced by diamond wire saws is comparable to those sliced by slurry wiresaws, after the application of surface etching and stress relaxation, as discussed in Section 6.4.3.

6.4.5   Cost of Wafering

Moller (2006) reported that the cost of operating slurry wire saws accounts for approximately 30% of the total cost of solar wafer production, enough to drive most manufacturers away from slurry wiresaws towards diamond wire saws. However, Kray et al. (2007) compared the performance and cost of operating slurry wiresaws and diamond wire saws, and demonstrated that diamond-impregnated wires are so expensive that there was no benefit to replacing slurry wiresaws with diamond wire saws, unless the cost of diamond wire could be reduced by a factor of three. Despite these results, studies have demonstrated several advantages of using diamond wire saws. These considerations include longer wire life, lower costs for the cutting fluid, the ability to recycle material from kerf loss [Goodrich et al. (2013)], shorter machining times, and the simplicity of the slicing fluid. The slicing fluid used in diamond wire saws is a simple aqueous surfactant solution, and is cheaper and easier to remove than polyethylene glycol – a typical carrier fluid of slurry for the slurry wiresaws, consisting of carrier fluid and abrasive grits (cf. Section 3.8). Developments in cost reduction of diamond wire and operations have made diamond wire saws become widely used now. For example, Goodrich et al. (2013) predicted that the use of diamond wire saws could reduce the cost of solar wafers by USD11 per square meter[4]. It appears that diamond wire saws may be poised to replace slurry wiresaws for wafer slicing in the near future.

6.5 Slicing Performance with Different Process Parameters

The slicing parameters of diamond wire saws, such as wire speed, feed rate, and wire tension, affect the surface roughness of wafers. The size and distribution of diamond grains on the wire also play an important role in the outcomes of wafer properties. Experimental results [Chung and Nhat (2014); Teomete (2011b)] showed good agreement with computational models when a simplified model, assuming the sliced wafer surface is under ductile-regime machining, was used to study the surface roughness of as-sliced wafers [Chung and Nhat (2014)]. The effects of parameters on the surface roughness of wafers are summarized in the following sections.

6.5.1 Effect of Wire Speed

Both simulation [Chung and Nhat (2014)] and experimental [Teomete (2011b)] results show that the surface roughness of alumina ceramics sliced by diamond wire saws will be reduced when the speed of the diamond wire is increased. These results are similar to other metal machining processes, such as turning and milling. Usually, a higher cutting speed can reduce surface roughness unless the interfacial effects, such as heat or build-up edge, deteriorate the surface. However, the experimental results of the surface roughness of silicon carbide (SiC) wafers sliced by diamond wire saws did not follow this rule [Huang et al. (2015)]. The surface roughness did not show significant difference over the range of wire speed from 3.33 to 10 m s^{-1}. One possible reason could be that the extremely high hardness of SiC can limit the formation of ductile-regime machining. Further study is required.

6.5.2 Effect of Feed Rate

Unlike the wire speed discussed in Section 6.5.1, the increase of feed rate will have an adverse effect on the surface roughness of wafers sliced by diamond wire saws. That is, an increase in feed rate will increase the surface roughness, similar to many metal machining processes. With higher feed rates, abrasive grits on the diamond wire will be subject to more contact interaction with engagement to remove more material, causing more fractures and surface scratches, resulting in increased surface roughness. Although a better surface can be achieved by reducing the feed rate, the reduction of feed rate will also reduce the yield of production of wafers. This is a trade-off that must be determined and optimized by the manufacturer.

---

4 The total manufacturing cost of solar wafer is USD76.

However, experimental results have shown that the feed rate does not change the surface roughness significantly in the slicing of a SiC ingot [Hardin et al. (2004); Huang et al. (2015)].

6.5.3 Effect of Grain Density

The diamond grain density, which is the number of abrasive grains per unit circumferential area of the core wire, affects the number of actively engaged diamond grits during the machining process of operating diamond wire saws. Both simulation [Chung and Nhat (2014)] and experimental [Teomete (2011b)] results also show that higher grain density (with smaller spacing between the abrasives) results in a better surface finish. However, when the diamond grain density is too high, it will leave reduced space for the chips and cutting fluid around the wire that may affect the effectiveness of slicing. In addition, higher diamond grain density may increase the number of load-bearing abrasives, resulting in the decrease of load per each load-bearing abrasive grits and reduction of cutting efficiency.

6.5.4 Effect of Wire Tension

Many experimental results have shown that the wire tension does not affect the surface roughness of wafers significantly [Hardin et al. (2004); Huang et al. (2015); Teomete (2011b)]. The range of the tension maintained in these studies was from 13.3 to 35 Newtons. The bow angle formed during slicing depends on the material removal rate, feed rate, wire tension, workpiece material, etc. If other parameters were maintained at the same level, the effects of wire tension and feed rate could compromise with each other to reach a balanced bow angle in order not to affect the surface roughness [Teomete (2011b)].

In general, an increase in feed rate is necessary to increase productivity. However, this will also increase the surface roughness. One remedy to reduce the surface roughness is to increase the wire speed at the same time, because higher wire speed can reduce the surface roughness. With careful arrangement, increasing wire speed and feed rate can increase the productivity without further deteriorating the surface roughness, as long as the wire speed does not exceed the machine limit.

In addition, the diamond wire must be strong enough to survive under the conditions of high speed and cutting force. Diamond wire with larger diameter is better suited for a higher production rate when operating at higher speeds. However, diamond wire with larger diameter removes more material with larger kerf loss than a thinner wire. This can reduce the total number of wafers produced from ingot.

6.6 Summary

In this chapter, various aspects of diamond wire saws are presented, including the design and manufacturing of diamond wires, the slicing mechanism, comparison between slurry wiresaws and diamond wire saws, and slicing performance versus process parameters.

The design and manufacturing of diamond wires have made significant improvement over the last decade to make diamond wire saws a formidable machine tool for the slicing of wafers from ingots. It appears that diamond wire saws may be poised to replace slurry wiresaws with further technological improvements in the near future, just as slurry wiresaws did for ID saws in the mid to late 1990s. The advantages of diamond wire saws include easy cleaning, longer tool life, higher cutting efficiency, and lower cost of coolant or slurry. The challenges of diamond wire saws include the cost of making diamond wires and weak bonding of diamond grits onto the wire's circumferential surface. Significant improvements have been made in manufacturing diamond wires, making the diamond wire saw an increasing popular machine tool for wafer slicing in industry. However, issues such as scratch tool marks, different bi-directional fracture strengths, and more brutal brittle machining still require more research and study, especially for slicing polycrystalline PV wafers that are typically thinner.

References

Bidiville A, Wasmer K, Kraft R and Ballif C 2009 Diamond wire-sawn silicon wafers - from the lab to the cell production *Proceedings of the 24th EU PV-SEC*, pp. 1400–1450, Hamburg, Germany.

Chen CCA and Chao PH 2010 Surface texture analysis of fixed and free abrasive machining of silicon substrates for solar cells. *Advanced Materials Research* **126-128**, 177–180.

Chen W, Liu X, Li M, Yin C and Zhou L 2014 On the nature and removal of saw marks on diamond wire sawn multicrystalline silicon wafers. *Materials Science in Semiconductor Processing* **27**(1), 220–227.

Chiba Y, Tani Y, Enomoto T and Sato H 2003 Development of a high-speed manufacturing method for electroplated diamond wire tools. *CIRP Annals - Manufacturing Technology* **52**(1), 281–284.

Chung C and Nhat LV 2014 Generation of diamond wire sliced wafer surface based on the distribution of diamond grits. *International Journal of Precision Engineering and Manufacturing* **15**(5), 789–796.

Chung C and Nhat LV 2015 Depth of cut per abrasive in fixed diamond wire sawing. *The International Journal of Advanced Manufacturing Technology* **80**, 1337–1346.

Clark W, Shih A, Hardin C, Lemaster R and McSpadden S 2003a Fixed abrasive diamond wire machining- part i: Process monitoring and wire tension force. *International Journal of Machine Tools and Manufacture* **43**(5), 523–532.

Clark W, Shih A, Lemaster R and McSpadden S 2003b Fixed abrasive diamond wire machining- part ii: Experiments design and results. *International Journal of Machine Tools and Manufacture* **43**(5), 533–542.

Enomoto T, Shimazaki Y, Tani Y, Suzuki M and Kanda Y 1999 Development of a resinoid diamond wire containing metal powder for slicing a slicing ingot. *CIRP Annals - Manufacturing Technology* **48**(1), 273–276.

Ge PQ, Gao YF, Li SZ and Hou ZJ 2009 Study on electroplated diamond wire saw development and wire saw wear analysis. *J Engineering Materials* **416**, 311–315.

Goodrich A, Hacke P, Wang Q, Sopori B, Margolis R, James TL and Woodhouse M 2013 A wafer-based monocrystalline silicon photovoltaics road map: Utilizing known technology improvement opportunities for further reductions in manufacturing costs. *Solar Energy Materials and Solar Cells* **114**, 110–135.

Hardin CW, Qu J and Shih AJ 2004 Fixed abrasive diamond wire saw slicing of single-crystal silicon carbide wafers. *Materials and Manufacturing Processes* **19**(2), 355–367.

Huang H, Guo L and Xu X 2009 Experimental investigation of temperatures in diamond wire sawing granite. *Key Engineering Materials* **404**, 185–191.

Huang H, Zhang Y and Xu X 2015 Experimental investigation on the machining characteristics of single-crystal sic sawing with the fixed diamond wire. *International Journal of Advanced Manufacturing Technology* **81**, 955–965.

Ide D 2007 Resin bond diamond wire for slicing ceramics. *Industrial Diamond Review* **2**, 32–34.

Kray D, Schumann M, Eyer A, Willeke G, Kubler R, Beinert J and Kleer G 2007 Solar wafer slicing with loose and fixed grains *Conference Record of the 2006 IEEE 4th World Conference on Photovoltaic Energy Conversion, WCPEC-4*, vol. 1, pp. 948–951, Waikoloa, HI, United states.

Lippold M, Buchholz F, Gondek C, Honeit F, Wefringhaus E and Kroke E 2014 Texturing of sic-slurry and diamond wire sawn silicon wafers by HF-HNO 3-H2SO4 mixtures. *Solar Energy Materials and Solar Cells* **127**, 104–110.

Meng J, Li J, Ge P and Zhou R 2006 Research on endless wire saw cutting of al2o3/tic ceramics, vol. **315-316**, pp. 571–574.

Moller H 2006 Photovoltaics - current status and perspectives. *Environment Protection Engineering* **32**(1), 127–134.

Pauli P, Beesley JG, Schonholzer UP and Kerat U 2005 Swiss wafer slicing technology for the global PV market from Meyer + Burger AG - novel trends for the future in photovoltaic wafer manufacturing *6 Symposium Photovoltaic National SIG Geneve*.

Shimazaki Y, Enomoto T and Tani Y 2002 Abrasive wire for a wire saw and a method of manufacturing the abrasive wire. US Patent 6,463,921.

Taniguchi K, Nakano M and Manita Y 2010 Electrodeposited wire tool. US Patent 7,704,127.

Teomete E 2011a Investigation of long waviness induced by the wire saw process. *Proceedings of the Institution of Mechanical Engineers, Part B: Journal of Engineering Manufacture* **225**(7), 1153–1162.

Teomete E 2011b Roughness damage evolution due to wire saw process. *International Journal of Precision Engineering and Manufacturing* **12**(6), 941–947.

Ueoka I, Sugawara J, Mizoguchi A, Oshita H, Yamanaka M, Ogawa H, Urakawa N and Yoshinaga H 2000 Wire-saw and its manufacturing method. US Patent 6,070,570.

Watanabe N, Kondo Y, Ide D, Matsuki T, Takato H and Sakata I 2010 Characterization of polycrystalline silicon wafers for solar cells sliced with novel fixed-abrasive wire. *Progress in Photovoltaics: Research and Applications* **18**(7), 485–490.

Wu H and Melkote SN 2012 Study of ductile-to-brittle transition in single grit diamond scribing of silicon: Application to wire sawing of silicon wafers. *Journal of Engineering Materials and Technology, Transactions of the ASME*.

Wu H, Melkote SN and Danyluk S 2012 Mechanical strength of silicon wafers cut by loose abrasive slurry and fixed abrasive diamond wire sawing. *Advanced Engineering Materials* **14**(5), 342–348.

Wu H, Yang C and Melkote SN 2014 Effect of reciprocating wire slurry sawing on surface quality and mechanical strength of as-cut solar silicon wafers. *Precision Engineering* **38**(1), 121–126.

Yang C, Mess F, Skenes K, Melkote S and Danyluk S 2013a On the residual stress and fracture strength of crystalline silicon wafers. *Applied Physics Letters.*

Yang C, Wu H, Melkote S and Danyluk S 2013b Comparative analysis of fracture strength of slurry and diamond wire sawn multicrystalline silicon solar wafers. *Advanced Engineering Materials* **15**(5), 358–365.

Part III

Wafer Surface Preparation and Management

7

Lapping

7.1 Introduction

After wafers are sliced from crystal ingots, individual as-sliced wafers have tool marks on the surface, as well as subsurface cracks and damage beneath the surface. The next steps in wafer surface processing involves the removal of tool marks and the flattening of the wafer surface. Figure 7.1 illustrates the "polishing" processes that include lapping, etching, pre-polishing, and polishing to produce polished wafers, ready for microelectronic fabrication. The focus of this chapter is the "lapping" process.

Typical processes to flatten wafer surfaces include grinding or lapping. The processes can also be broken into single-sided or double-sided operation. The determining factors in adopting lapping or grinding, or both, are (i) wafer size, (ii) wafer material, and (iii) type of device to be fabricated. The lapping process is the most economical way to perform thinning or flattening of silicon wafers of size 200 mm or smaller. Lapping is a step to achieve a high degree of flatness of wafers, and to reduce the depth of subsurface damage, and radial and lateral cracks formed during the slicing process. Since lapping is also a mechanical process with free abrasive machining, new cracks and damage beneath the surface will be generated, although with a shallower depth than those produced by more brutal slicing processes. Wafers produced by industrial processes typically are lapped on both sides.

The main objectives of lapping are to obtain a high level of uniformity, parallelism and flatness of wafer surfaces, and to achieve a more precise thickness of wafers according to their sizes (see also Section 1.3.1). Furthermore, lapping is a process of removing primary surface and subsurface damage, and thinning the wafer. The sizes of wafers that can be effectively processed by lapping are typically 200 mm or smaller,

Wafer Manufacturing: Shaping of Single Crystal Silicon Wafers,
First Edition. Imin Kao and Chunhui Chung.
© 2021 John Wiley & Sons Ltd. Published 2021 by John Wiley & Sons Ltd.

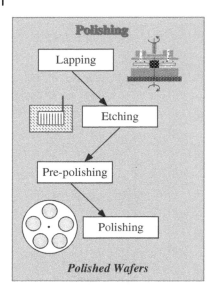

Figure 7.1 Processes of "wafer polishing" which includes lapping, etching, and polishing in wafer manufacturing (see also Figure 2.1 in Chapter 2).

although wafers of larger size also can also be processes by the lapping process. The lapping process can accomplish several things, including the following:

- It removes tool marks (grooves or trenches/cracks produced by various types of wire saws and/or by ID saws) and surface defects from the both side of the wafers.
- It reduces the thickness of the wafer and relieves the residual stress resulting from the slicing process.
- Before and after the lapping process, wafers are subjected to many in-process inspections to evaluate their quality, flatness, and planarization.

7.2 Fundamentals of Lapping and FAM

Lapping is a free abrasive machining (FAM) process, as presented in Section 3.4.2. In such FAM processes, the abrasive grits are free to roll or move in a "three-body abrasion," as illustrated in Figure 3.6. The "three bodies" are the tool, the workpiece surface, and the abrasive grits as three distinct entities, with motion relative to one another, without physical bonding between them. The machining from abrasion is based on the relative motion between the three bodies, while the abrasive grits roll, indent, and scratch the surface of the workpiece, as well as the tool. Therefore, the FAM processes are broken into two mechanisms and their combinations: (i) rolling-indenting mechanism, (ii) rolling-scratching mechanism, and (iii) a combination of rolling-indenting and rolling-scratching. Figure 5.4 illustrates the rolling-indenting and rolling-scratching processes and the interplay of both mechanisms in most FAM processes. In general FAM processes, both rolling-indenting and scratching-indenting processes can take place with different degrees of engagement, as illustrated in Figure 5.4. The composition and percentage

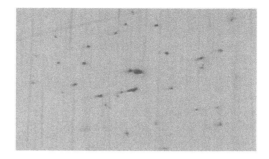

Figure 7.2 Surface of a silicon wafer after a lapping operation with larger abrasive grit size, showing scratches on the surface. This indicates that the larger abrasive grits used in this lapping process are directly pressed unto the wafer surface by the lapping tool, causing rolling and scratching abrasion on the surface. Source: Imin Kao.

of engagement of both processes depends on the nature of FAM activities at the contact interface of machining.

A lapping process is typical dominated by the rolling-scratching FAM process. Figure 7.2 shows a surface of silicon wafer after a lapping operation using larger abrasive grits. Visible scratching tool marks can be seen on the surface owing to the rolling-scratching abrasion. Such visible scratching tool marks may be reduced with smaller abrasive grits, as well as with well controlled manufacturing process parameters. A well designed lapping process can render a surface that is free of scratches.

Lapping of wafers, due to the nature of its mechanical abrasion and FAM process, provides the following outcomes:

1. Flattens the surface of wafers after they are sliced.
2. Reduces the waviness and thickness variation of the wafers.
3. Achieves preliminary planarization of the wafer surface.
4. Removes a layer of subsurface cracks, created by the mechanical slicing processes. Although also a mechanical abrasion process, lapping generates a lower degree of subsurface damage. The lapping process is normally well controlled such that the shallow layer of subsurface damage due to the lapping operation can be readily removed in the subsequent polishing process.
5. Relieves the residual stress on wafers, created by the slicing operations, through the mechanical process of contact and abrasion. This is also called "stress relaxation."

Lapping of silicon wafers is performed using lapping plates (for example, cast iron), an abrasive material (for example, aluminum oxide), and a lapping slurry additive. The lapping slurry additive suspends the mixture of abrasive grits and water, creating an effective slurry to sustain the lapping operation.

Lapping slurry can be broken into two main types:

(i) Oil-based lapping slurry
(ii) Water-based lapping slurry.

The slurry is supplied through a non-shear pump to the top of the lapping equipment and is dispersed to the interface between the silicon wafers and lapping plates (see also Figure 7.5). The lapping slurry should be stable in supply and composition for at least an eight-hour shift in industrial applications. The lapping slurry must be able to:

- Sustain slurry film strength to separate the wafer from direct contact with the lapping plates in order to keep the wafers from breaking.
- Provide corrosion protection to keep the plate from rusting.
- Maintain detergency to keep the wafers from staining when they are removed from the lapping equipment after the lapping operation is finished.

Types of Lapping Methods

As the diameter of wafers has increased over the decades, lapping methods have evolved to respond to the needs of wafer preparation. Two types of lapping methods are

 (i) Single-sided lapping
(ii) Double-sided lapping.

The adoption of single-sided or double-sided lapping depends on the different needs and applications of wafers. More details will be presented in Section 7.3.

Lapping Abrasive and Slurry

Lapping slurry for FAM equipment consists of an abrasive material, a carrier fluid (water or oil), and slurry additive. The carrier and slurry additive suspend free abrasive grits and will keep the wafer from breakage. Mixing ratios for the lapping slurry can range from 1:10 to 1:5.[1] When water is used as the carrier fluid, de-ionized water is often used in both single-sided and double-sided lapping.

The lapping abrasives usually are silicon carbide, aluminum oxide, diamond or others, although aluminum oxide (sometimes called *alumina*) is the most commonly used abrasive for lapping. The abrasive grit sizes range from 17 to 8 microns.[2]

The choice of abrasive depends on the type of material; for example, a very aggressive abrasive, such as diamond, will cause deeper subsurface damage. The subsurface damage can be reduced by optimizing the lapping process parameters. As an example, the subsurface damage of wafer surface can be reduced by decreasing the load on the wafers and reducing the plate speed as the final thickness is approached.

Lapping Versus Grinding Followed by Lapping

Lapping has been the conventional method of wafer surface processing to obtain a high degree of wafer flatness through a simple and easy-to-use methodology. The lapping operation is employed in wafer surface processing for many materials, such as silicon, silicon carbide, aluminum oxide, sapphire, lithium niobate, I–VII, III–V, III–VI, IV–VI compounds, and other materials. Lapping can sustain operation of wafer surface processing to obtain minimal thickness variation and stress-free surfaces as the wafer size increases from 100 to 200 mm and to 300 mm.

---

1 The mixing ratio is calculated using a kilogram of abrasive grits versus a liter of carrier fluid. Thus, the 1:10 ratio uses 0.1 kg (or 100 g) of abrasive grits to mix with 1 L of carrier fluid. When using water and anticorrosion additives, the mixing ratio can be as high as 1:3 to 1:2 (333–500 g of abrasives per liter of carrier fluid.) See also Sections 3.8 and 3.8.1.
2 The range of abrasive grit size corresponds to 400, 500, and 600. See also Section 3.5 and Tables 3.1 and 3.2 for more detail on abrasive grit and microgrit size.

The lapping operation has been expanded to handle 300 mm silicon wafers in wafer manufacturing facilities since the 300 mm silicon wafers were introduced. With larger wafer surface[3], some have suggested that a combination of surface grinding and lapping was best. The combination employs first a diamond grinding process on the surface of 300 mm wafers, followed by lapping to finish the surface processing. However, others argue that only lapping can consistently produce flat, stress-free (by relieving residual stress), low damage wafers after slicing operation economically and efficiently. It appears that the choice is a preference of the individual wafer processing company, although the dispute is still ongoing.

7.3 Various Configurations and Types of Lapping Operation

Single wafer lapping equipment, the Logitech PM5 lapping equipment, is shown in Figure 7.3, which is often employed in labs to conduct research studies in lapping. Although with simple construction for single-sided wafer lapping, such single wafer lapping equipment is very useful in illustrating the fundamental configuration and operation of lapping. The main components of a lapping machine include:

- Lapping plate: The lapping plate provides the support for the wafer mounted on the jig (or held by a clamping device), and performs the function of lapping. The lapping plate can be made of cast iron with grooves to facilitate the flow of slurry and silicon kerf from the lapping operation. Different groove designs include: radial grooves, waffle pattern grooves, spiral grooves, and non-groove. Lapping plate can also be coated with other soft pads, such as felt, to perform lapping with soft pads. When double-sided lapping is desired, two lapping plates need to be employed on both the top and bottom sides of the wafer.

Figure 7.3 Logitech PM5 lapping equipment for conducting lapping research at a laboratory. Source: Imin Kao.

---

3 The surface area increases by 125% (more than doubled) when the diameter of the wafer increases from 200 to 300 mm.

- A clamping device that holds the wafer: A jig or a clamping device or mechanism is used to hold the wafer in place and maintain contact with the surface of the lapping plate. A prescribed normal force at the contact is maintained. The jig or clamping device can also have an indicator gauge or sensor to measure the depth of lapping.
- Slurry and slurry feeding mechanism: Slurry is mixed and fed to the lapping interface between the wafer and the lapping plate to perform the FAM. For consistency and sustained lapping operation, a slurry tank or equivalent feeding mechanism must be provided.
- Others: A means of fixing the wafer to the jig, or a clamping device, is needed for single-sided lapping, for example, using bees wax or epoxy. It is noted that the wafer and the lapping plate are typically rotating in opposite directions to each other to enhance the machining operation.

Figure 7.3 shows a Logitech PM5 lapping equipment for single-sided lapping. It shows a jig assembly holding a wafer of 100 mm diameter, a cast iron lapping plate on which the jig assembly and wafer rest, a slurry tank, and a slurry dispenser. The jig assembly holding the wafer is configured to rotate in the opposite direction to the direction of the lapping plate. The slurry tank has a motorized system to constantly rotate and mix the slurry in the tank, while feeding the well mixed slurry to the interface between the wafer and the lapping plate. A control panel is also provided to set up various parameters of the lapping operation.

A Schematic of Single-sided Lapping Equipment
A schematic of a single-sided lapping equipment for wafers is shown in Figure 7.4. The upper part of the schematic illustrates a spindle assembly with a wafer holder at the bottom of the assembly to house the wafer to be lapped and to hold it in place while performing the lapping operation. The wafer can be clamped or fixed or glued to the spindle assembly. The bottom plate in Figure 7.4 is a lapping plate, typically

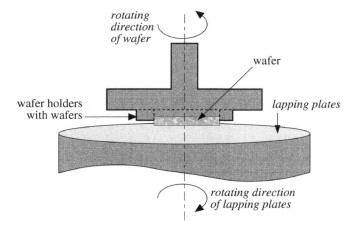

Figure 7.4 A schematic of a single-sided lapping machine.

made of cast iron or similar materials. The lapping plate also has grooves to facilitate the flow of slurry and silicon kerf. A soft pad can be placed on the lapping plate to provide soft-pad lapping. Although not shown in the figure, slurry is dispensed to the lapping plate to provide the FAM action at the contact interface between the held wafer and the lapping plate. It is important to note that the top spindle assembly and the bottom lapping plate rotate in opposite directions to optimize the lapping performance.

A Schematic of Double-sided Lapping Equipment

A schematic of an industrial lapping machine is illustrated in Figure 7.5 which is typically configured to perform double-sided lapping. A similar set up can be configured to perform single-sided lapping operation. The schematic of the front view illustrated in Figure 7.5(top) shows two lapping plates for double-sided lapping: one on the

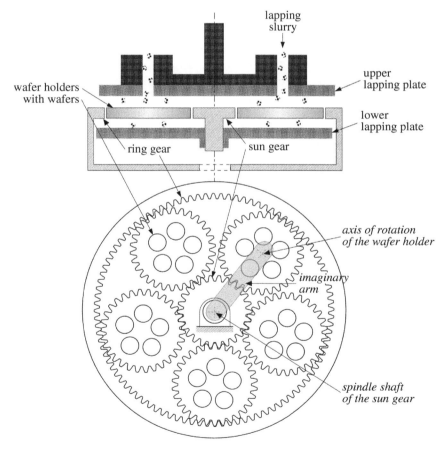

Figure 7.5 A schematic of an industrial double-sided lapping machine with five wafer holders, each holding several wafers: (top) a schematic of the front view of typical industrial lapping machine, and (bottom) a schematic of the top view of the lapping machine with a planetary gear train.

top, called the upper lapping plate, and the other at the bottom, called the lower lapping plate. The upper lapping plate is always floating rather than rigid coupling. The lapping slurry is fed from the supply through the holes on the upper lapping plate to evenly distribute in the work zone when the lapping operation is in progress to ensure that both surfaces can be lapped at the same time to produce similar surface profile on both surfaces. The rotation speed is typically lower than 100 revolutions per minute (RPM). In some cases, in situ thickness and flatness measurement are built into the lapping machine. Several wafers are placed in wafer holders, which have gear teeth to mesh with the sun gear in the middle and the ring year on the outer ring, as illustrated in the top view of the lapping machine in Figure 7.5.

Analysis of the Planetary Gear Train of the Double-sided Lapping Equipment
The sun gear, wafer holders (having gear teeth on the circumferential exterior), and the ring gear form the arrangement of a planetary (or *epicyclic*) gear train, as illustrated in Figure 7.5. For the analysis of the planetary gear train, we introduce an imaginary arm connecting and holding together the spindle shaft of the sun gear and the axis of rotation of the wafer holder. The imaginary arm of the planetary gear train is illustrated in Figure 7.5(bottom) connecting the spindle shaft of the sun gear and the axis of rotation of the wafer holder. The imaginary arm keeps the wafer holder gear meshing with the sun gear and causes it to revolve around the sun gear, while the wafer holder gear meshes with the ring gear when the planetary gear train is set in motion. Such an imaginary arm already exists in the planetary gear train of the lapping equipment because the sun gear, wafer holder gear, and the ring gear are working together on the same plane and meshing with one another. The upper lapping plate holds the entire planetary gear assembly in planar motion, while the lower lapping plate supports the entire planetary gear assembly. Now that we have the configuration of the planetary gear train, we are ready to proceed with the kinematic analysis of motion.

Assume the speeds of the the sun gear, the imaginary arm, and the ring gear are ω_s, ω_A, and ω_r, respectively. The equation for the analysis of the planetary gear train is[4]:

$$e = \frac{\omega_r - \omega_A}{\omega_s - \omega_A} \tag{7.1}$$

where e is the gear train ratio from the sun gear to the ring gear, following the convention of the sign of direction of motion assuming direct spur gear contact. The rotation speed of the gears also follows the convention of positive sign for counterclockwise rotation, and negative sign for clockwise rotation. The train ratio, therefore, is

$$e = \left(-\frac{N_s}{N_w}\right)\left(+\frac{N_w}{N_r}\right) \tag{7.2}$$

where N_s, N_w and N_r are the number of teeth of the sun gear, the wafer holder, and the ring gear, respectively. The first gear ratio has a negative sign because the direction of rotation of the sun gear and wafer holder are in opposite directions,

---

4 The readers may refer to Arnaudov and Arnaudov (2019); Norton (2020); Shigley and Mischke (2002), for example, for further reference.

assuming a direct spur gear contact between them. The second gear ratio has a positive sign because the direction of rotation of the wafer holder and the ring gear are in the same direction, assuming a direct spur gear contact between them. Note that $N_r = N_s + 2 N_w$. Put Equations (7.1) and (7.2) together to render

$$\left(-\frac{N_s}{N_r}\right) = \frac{\omega_r - \omega_A}{\omega_s - \omega_A}. \qquad (7.3)$$

Let's look at an example in the following to illustrate the application of the equations for the analysis of the planetary gear train in a double-sided lapping machine.

Example 7.3.1

The planetary gear train of the double-sided lapping equipment, shown in Figure 7.5(bottom), has the following number of gear teeth: $N_s = 80^T, N_w = 160^T$, and $N_r = 400^T$. Assume the ring gear is held stationary with $\omega_r = 0$. The sun gear has a speed of $\omega_s = 60$ RPM. Determine the speed of rotation for the wafer holder.

Solution

Assume an imaginary arm connecting the axis of rotation of the sun gear and the axis of rotation of the wafer holder, with a speed of ω_A. The train ratio in Equation (7.2) is

$$e = \left(-\frac{80}{160}\right)\left(\frac{160}{400}\right) = \left(-\frac{1}{5}\right).$$

Employing Equations (7.2) and (7.3) using the given parameters, we have

$$\left(-\frac{1}{5}\right) = \frac{0 - \omega_A}{60 - \omega_A}.$$

Solve for $\omega_A = 10$ RPM. This means that the imaginary arm connecting the sun gear and wafer holder rotates at 10 RPM in the counterclockwise direction. Next, we can apply the same Equation (7.2) using the train ratio from the sun gear to the wafer holder, with the corresponding gear train, as follows

$$\left(-\frac{N_s}{N_w}\right) = \frac{\omega_w - \omega_A}{\omega_s - \omega_A} \implies \left(-\frac{80}{160}\right) = \frac{\omega_w - 10}{60 - 10} \implies \omega_w = -15.$$

From the equation, the speed of the wafer holder is found to be $\omega_w = -15$ RPM. The negative sign of ω_w indicates that the wafer holder rotates in the clockwise direction, opposite to the direction of rotation of the sun gear.

Note that the wafer holder in Example 7.3.1 rotates at a speed of 15 RMP in the clockwise direction. This is in the opposite direction to the direction of rotation of the sun gear that carries the lapping plate with wafers. It does not, however, mean that all wafers held in the wafer holder will rotate at exactly the same speed as that of the wafer holder because each wafer is allowed to have relative rotation with respect to the wafer holder. This is to allow the freedom for the wafer to have stress-free motion throughout the lapping process.

Equations (7.1)–(7.3) can be applied to the analysis of different speed arrangement of the planetary gear train, following the same procedures of analysis as those in Example 7.3.1. These equations can be also employed to synthesize the desired arrangement for a specific speed of the wafer holder. In the following Example 7.3.2, a synthesis is provided to determine the required speed of the sun gear in order to meet the requirement of a desired rotational speed of the wafer holders.

Example 7.3.2
Example 7.3.1 is an analysis of the planetary gear train of a double-sided lapping machine, with the prescribed angular speeds of the sun gear and ring gear. Use the same parameters in Example 7.3.1 to synthesize and design the speed of the sun gear to achieve a speed of rotation for the wafer holder at 45 RPM, in an opposite direction of rotation to the sun gear.

Solution
We will use the same number of gear teeth for the gear train in Example 7.3.1, with a stationary ring gear. The wafer holders are assumed to rotation in the clockwise sense with a speed of $\omega_w = -45$ RPM. The overall train ratio is $e = \left(-\frac{1}{5}\right)$. Equations (7.1)–(7.3) become

$$\left(-\frac{1}{5}\right) = \frac{0 - \omega_A}{\omega_s - \omega_A} \tag{7.4}$$

$$\left(-\frac{80}{160}\right) = \frac{-45 - \omega_A}{\omega_s - \omega_A} \tag{7.5}$$

where Equation (7.4) describes the planetary gear train from the sun gear to the ring gear, and Equation (7.5) describes the planetary gear train from the sun gear to the wafer holder. Divide Equation (7.4) by Equation (7.5) to obtain

$$\frac{2}{5} = \frac{-\omega_A}{-45 - \omega_A}.$$

Solve for $\omega_A = 30$ RPM. This means that the arm rotates in the counterclockwise direction at a speed of 30 RPM. Substituting ω_A into Equation (7.4), we can write

$$\left(-\frac{1}{5}\right) = \frac{-\omega_A}{\omega_s - \omega_A} = \frac{-30}{\omega_s - 30}.$$

Solve for $\omega_s = 180$ RPM. Therefore, the sun gear must rotate at a speed of 180 RPM in the counterclockwise direction in order to produce the desired speed of the wafer holders at 45 RPM in the clockwise direction. In addition, the rotation of the sun gear and the wafer holder are in opposite directions, as required.

7.3.1 Single-sided Lapping

Single-sided lapping was discussed in Section 7.3 earlier with an illustration in Figure 7.4. An industrial single-sided lapping machine is a viable alternative to surface grinding operations and double disk grinding, while providing a good surface finish. However, it is time consuming if both sides of the wafer surface must be lapped by employing single-sided lapping equipment. Silicon wafers today require lapping on both sides, making the single-sided lapping machine not a favorable choice in mass production [Barylski and Piotrowski (2017); Doi et al. (2012)]. This is especially true for wafers of large diameter, such as the 300 mm silicon wafers.

7.3.2 Double-sided Lapping

Double-sided lapping provides simultaneous lapping on both sides of the wafers. Figure 7.5 in the preceding presentation of this section provides technical discussions of industrial lapping machines. Double-sided lapping can achieve an ultra smooth surface on silicon wafers, as well as other hard and brittle materials, such as sapphire, silicon carbide, and other III-V and II-VI semiconductor materials. The industrial double-sided lapping machines can provide simultaneous lapping of multiple wafers on both sides to help reduce the cost and time of lapping operation.

Double-sided lapping can be broken into two types by the nature of abrasives, as follows.

- FAM lapping: The first type is a double-sided lapping with free abrasives, in a FAM process. See also Section 3.4.2. The abrasives used in this free abrasive lapping are normally aluminum oxide, although other abrasive materials are also used. See also the topic on lapping abrasive and slurry in Section 7.2.
- BAM lapping: The second type is a double-sided lapping with bonded abrasives. See also Section 3.4.1. This type of bonded abrasive lapping is often performed with bonded abrasive on soft textured pads, sometimes called soft-pad lapping.

The material removal rate (MRR) of the lapping processes can be controlled by different process parameters, such as size of abrasive grits, relative speed of lapping plates, normal load, slurry flow, and others. A well-controlled lapping process can achieve parallelism and planarization of wafer surfaces, discussed in Section 7.4, with very flat surface finish with few scratches and little subsurface damage.

With efficient processing of surfaces of 300-mm silicon wafers, double-sided grinding machine has been developed to complement the conventional batch processing using double-sided lapping machines. However, the double-sided grinding machine is typically design for single wafer processing, making it less efficient in comparison with the bath processing of multiple wafers on the industrial double-sided lapping machines.

7.3.3 Soft-pad Lapping

Soft-pad lapping has different embodiment. It can be a regular lapping machine with the replacement of cast iron lapping plate by a plate with soft pad. It can also be a BAM-type double-sided lapping operation that employs bonded abrasive, typically diamond grits, on a textured pad, as discussed earlier in this section on the topic of double-sided lapping. De-ionized water is typically used as the fluid for slurry in double-sided lapping with bonded abrasives. In BAM-type double-sided lapping, the textured pad dictates the trajectory of abrasive grits on the surface of silicon wafers, which may cause uniform tool marks, related to trajectory uniformity, during the material removal process. Such tool marks need to be removed with a subsequent process for better surface finish.

In addition to soft-pad lapping, a soft-pad grinding operation is developed for wafer surface processing. A perforated soft pad is inserted between the wafer and the ceramic chuck in the soft-pad grinding operation. The axis of rotation of the grinding wheel is offset by a distance of the wheel radius relative to the axis of rotation of the wafer.

7.3.4 Further References

For more reference reading materials on this topic, the readers may want to consult the following references, including, but not limited to, Cong et al. (2009); Fang et al. (2017); Fang et al. (2016); Guo (2019); Kasai (2008)' Klamecki (2001); Pan et al. (2019); Pan et al. (2012); Sahab et al. (2013); Wang et al. (2018); Wenski et al. (2002).

7.4 Lapping and Preliminary Planarization

Preliminary planarization of wafers accomplishes two purposes; one is to increase quality and parallelism of wafers and the other is to reduce the cost. Without global planarization of wafer surfaces, the thickness of wafer can be compromised and the time and cost of lapping and grinding can also be increased in order to reach the standards of the required wafer size and quality.

In the following, we discuss the needs for preliminary planarization of wafers and other relevant issues.

7.4.1 Quality Driven Needs for Preliminary Planarization

Prime wafers for the semiconductor fabrication of integrated circuits (IC) must meet rigorous surface flatness requirements. Lithographic processes[5] to print electronic IC designs via masks onto the wafer surface requires such prime wafers to be particularly flat in order to print IC designs on them (or on layers deposited upon them). The flatness or planarization of the wafer surface directly impacts device line width capability, process latitude, yield, and throughput. A high degree of planarization in the focal point of the electron beam delineator or optical printer is important for uniform imaging using lithography. The continuing reduction in device feature size [cf. Moore's law in Figure 1.4 of Section 1.3.3, and other references, such as Kulkarni and Desai (2001)] and increasingly stringent device fabrication specifications are forcing wafer manufacturing processes to adapt and prepare wafers with ever increasing flatness and planarization.

Wafer flatness measurements are presented in Section 1.4.4. Wafer flatness can be characterized in terms of typical terminology and characteristics using the four-character acronyms, as tabulated in Table 1.5:

1. Global flatness measurements: such as GBIR, or the TTV, GF3R, GF3D, GFLR, and GFLD. For example, GBIR, frequently used to measure global flatness variation, is the difference between the maximum and minimum thicknesses of the

---

5 For example, electron beam lithography or photolithography

wafer and is the same as the TTV. GBIR of a wafer is historically an important parameter of the quality of the wafer, with a reference plane being the ideal back surface on the entire fixed quality area (FQA), and flatness measurement in the range (TIR) between the maximum positive and negative deviations of global wafer surface.

2. Local site flatness measurements: such as SBIR, SBID, SF3R, SF3D, SFLR, SFLD, SFQR, and SFQD. For example, SBIR, frequently used to measure local site flatness variation, is the sum of the maximum positive and negative deviations of the surface in a small site of the wafer from an ideal back surface reference plane.
3. Subsite flatness measurements: such as SFQR and SFQD. For example, SFQR, which is becoming more widely used to measure the local site flatness range, is the sum of the maximum positive and negative deviations of the surface in a small site of the wafer with respect to a least-squares best fit reference plane.

A detailed discussion on the wafer flatness can be found in Section 1.4.4. In addition, readers may want to consult other reference of the SEMI document (https:// www.semiviews.org), ASTM document (www.astm.org) and others [Kulkarni and Desai (2001); Shimura (1988)].

7.4.2   Cost Driven Needs for Preliminary Planarization

Lapping is a machining process to flatten and reduce the roughness of wafer surfaces after the wafer is sliced from an ingot, as discussed in Section 7.2. Other than the initial capital cost of the equipment, the cost of lapping is primarily the consumables, including the slurry. Two "cost downs" of lapping are:

1. Reduction of the percentage of abrasive used in the lapping process for "cost down"
2. Recycling or recirculating abrasive for "cost down."

Other cost-driven approaches include the combination of grinding and lapping. As a case in point, Vandamme et al. (2000) in a US patent presented a method of processing as-sliced wafer surfaces by rough-grinding the front and back surfaces of wafers to quickly reduce the thickness of wafers. The front and back surfaces are then lapped with a lapping slurry to further reduce the thickness of the wafer and reduce damage caused by the mechanical grinding process. Such a technique of additional grinding before lapping can help decrease time, cost and size of wafer. Without rough grinding, a wafer may need 40 min to reduced its size by 80 microns by going though lapping only. With rough grinding on both the front and back of wafer surfaces, followed by lapping, it only takes 11 min to reduced the size by 80 micron.

However, some practitioners in the lapping industry still regard lapping as the best technology to produce consistently flat, stress-free, low damage wafers of various sizes from 100, 150, 200, 300mm and beyond economically and efficiently [Intersurface Dynamics (2019)], even though "it was once thought that a combination of surface grinding and lapping was best."

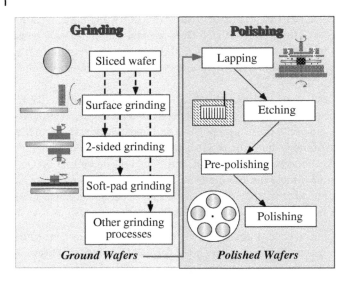

Figure 7.6 Schematic to illustrate the processes of "wafer grinding" followed by "wafer polishing" in wafer manufacturing, especially larger wafers (see also Figure 2.1 in Chapter 2),

Figure 7.6 illustrates schematically the wafer manufacturing process that includes grinding before the lapping and polishing processes. The grinding processes in Figure 7.6 include single-sided, double-sided, soft-pad, and other grinding processes.

Nevertheless, the global planarization of wafer surfaces can be achieved by combining rough grinding, fine grinding, and lapping, while optimizing time and cost of surface processing. This is especially important for larger wafers, such as silicon wafers of 300 mm diameter or larger that will benefit the subsequent polishing processes, including etching, polishing, and chemical-mechanical polishing (CMP), as illustrated in Figure 7.6, to produce prime wafers with the prescribed final wafer thickness.

7.5 Technical Challenges and Advances in Lapping

The lapping process enables wafers to reach a high degree of surface flatness, to produce parallelism between two surfaces, to reduce thickness variation, waviness and roughness, and to remove subsurface damage created in the slicing process. A proven process for small and large wafers of various materials, the lapping process continue to be the preferred methodology for wafer surface processing. Typically only a small layer of material is removed from the surface of wafers, ranging from as little as 5 μm to as much as 100 μm. Effective lapping processes require technical consideration and optimization. In the following, technical considerations and challenges are summarized with explanations [Dobrescu et al. (2014); Doi et al. (2012); Guo (2019); Marinescu et al. (2006); Sahab et al. (2013); Whitehouse (2010); Wikipedia (2018)].

- Abrasive materials: Effective material removal depends on the properties of the abrasive materials used in lapping. The material properties include grit size, grit integrity, geometry of the abrasives, hardness, and other mechanical properties. Refer to Section 3.5 in Chapter 3 for more detail.

- Abrasive sizes: The size of abrasive grits is one of the first considerations in lapping operations. Smaller abrasive grits tend to produce a wafer surface that is free of scratches and subsurface damage; however, it may require longer process time for meeting the requirements of material removal. A combination of abrasives to optimize the process time while producing a good surface can result in more effective time management of the lapping process and produce good surface finish.

- Pressure or load and speed: Lapping pressure has a significant effect on the efficiency of lapping, and will directly reflect in the final surface quality. The normal load is managed to maintain good contact in the free abrasive machining (FAM) process of lapping while ensuring that the wafers will not break. Therefore, understanding the FAM process and controlling of the process parameters are important. Refer to Sections 3.4.2 and 3.4.1. The speed of lapping is typically maintained at a speed lower than 100 RPM.

- Time of processing: Time of processing lapping is also an issue of mass production and cost. Industrial lapping is always a batch loading and machining process in order to increase the throughput and efficiency of the production of lapped wafers.

- Carrier fluid and slurry: Various studies have been perform in engineering and process modeling to understand the slurry characterization and optimization to produce more efficient lapping machines and to operate the FAM lapping processes. The slurry is typically a milk-like homogeneous mixture of the abrasive grits and carrier fluid (cf. Section 3.8.1). In lapping operations, carrier fluid can be oil-based, water-based, or water-soluble. Slurry is a consumable item containing both abrasive grits and carrier fluid. Appropriate management of slurry is important for efficiency and cost optimization, as well as for the minimization of environmental impacts.

- Others: In most cases, lapping is employed to process flat surfaces, such as crystalline wafers, not spherical shapes or contours. Specific lapping and polishing machines are designed for lenses with contours using different technologies, such as electro-rheological or magneto-rheological controlled surfaces. Lapping in wafer manufacturing is almost always a "wet process" that requires proper design considerations. The surface characteristics of lapped wafers are measured through periodic interruptions during the operating cycle, which can slow down the operation. A well controlled machining process with statistical process control (SPC), for example, can ensure the consistent outcome of the process.

7.5.1 Technical Considerations

Semiconductor materials are brittle. An important technical consideration is the strength of the wafer after back lapping[6]. During the thinning process using back

---

6 Back lapping, or back thinning processes in general, refers to a process that thins wafers prior to dicing to allow easy dice-sawing and to make the final thickness of electronic package to be

lapping, or similar processes, the machining process must avoid both surface and subsurface damage to the wafers. If the final thickness it too thin, making wafer too fragile to handle or be manipulated, it must be supported on a substrate, which can be another thicker wafer or a glass or ceramic substrate. The substrate should have the desired mechanical properties to provide necessary support, typically with a thickness-to-diameter ratio (or the aspect ratio) of at least 1:7. If such support is required, the second consideration is whether to make a permanent or temporary bond (cf. [SEMI 3D16 (2016); SEMI 3D2 (2013); SEMI 3D4 (2015); SEMI 3D8 (2014)]). This choice depends on the future use of the wafer and the required processing.

Grinding of as-sliced wafers before lapping has been suggested, as discussed earlier. The grinding of wafer surfaces provides certain advantages and reduction of time in surface processing. However, future design and development may provide the ability to process multiple wafers at the same time in an efficient and consistent process, with similar final outcomes of wafer surface characteristics as the lapping processes. In addition, the consistent patterns of tool marks on the ground wafers make them prone to fracture, and must be avoided. Until then, the double-sided batch industrial lapping machine is still the most economical machining operation for wafer surface processing that provides overall cost reduction of batch operations and the benefit of a proven machine tool and process.

7.5.2 Advances in Lapping

Advances in lapping include engineering, modeling, process understanding, slurry characterization and optimization, in situ measurement of wafer thickness, automatic plate flatness control, and back lapping in the thinning of wafers by removing material from the back surface, or the unpolished surface. Lapping processes and machines have continued to improve to respond to the evolving needs of industrial processing of wafer surfaces.

7.6 Summary

In this chapter, we start with the fundamental theory of lapping and free abrasive machining (FAM). Following that, various configurations and types of lapping machining are introduced, including single-sided, double-sided, and soft-pad lapping. Schematics of lapping machines are introduced, as well as typical industrial double-sided lapping machine and its planetary gear train, with analysis and synthesis of the angular speeds of the spindle axis and the wafer holders. The role of lapping in preliminary planarization of wafers is discussed in two aspects: one in quality-driven needs, and the other in cost-driven needs. Finally, some technical challenges and advances in lapping are discussed.

---

minimized. Thin wafers and IC dices improve the ability for heat transfer when the IC device must operate at high power level that generates a lot of heat. The heat transfer rate is inversely proportional to the size scale.

References

Arnaudov K and Arnaudov DP 2019 *Planetary Gear Trains* first edn. CRC Press.

Barylski A and Piotrowski N 2017 *Optimization of Single-Sided Lapping Kinematics Based on Statistical Analysis of Abrasive Particles Trajectories* IntechOpen; DOI: 10.5772/intechopen.71415.

Cong WL, Zhang PF and Pei ZJ 2009 Experimental investigations on material removal rate and surface roughness in lapping of substrate wafers: A literature review. *Key Engineering Materials* **404**, 23–31.

Dobrescu T, Pascu NE, Jiga G and Opran C 2014 Optimization criteria of plane lapping machines *25th DAAAM International Symposium on Intelligent Manufacturing and Automation*.

Doi T, Marinescu ID and Kurokawa S 2012 *Advances in CMP Polishing Technologies* Elsevier chapter 3 The Current Situation in Ultra-Precision Technology – Silicon Single Crystals as an Example, pp. 15–111.

Fang C, Zhao Z and Hu Z 2017 Pattern optimization for phyllotactic fixed abrasive pads based on the trajectory method. *IEEE Transactions on Semiconductor Manufacturing* **30**, 78–85.

Fang C, Zhao Z, Lu L and Lin Y 2016 Influence of fixed abrasive configuration on the polishing process of silicon wafers. *The International Journal of Advanced Manufacturing Technology* **88**, 575–584.

Guo J 2019 *Double-sided lapping, Encyclopedia* 1 edn. Scholarly Community Encyclopedia https://encyclopedia.pub/213.

Intersurface Dynamics 2019 Lapping Website. URL http://www.isurface.com/Catg2.asp?Cat2ID=55&Cat1ID=68.

Kasai T 2008 A kinematic analysis of disk motion in a double sided polisher for chemical mechanical planarization (CMP). *Tribology International* **41**, 111–118.

Klamecki BE 2001 Comparison of material removal rate models and experimental results for the double-sided polishing process. *Journal of Materials Processing Technology*.

Kulkarni M and Desai A 2001 Silicon wafering process flow. Technical report, PlasmaSil, LLC. US Patent 6,294,469.

Marinescu ID, Uhlmann E and Doi T 2006 *Marinescu-et-al-Book-2006*. CRC Press.

Norton R 2020 *Design of Machinery: An introduction to the synthesis and analysis of mechanisms and machines* 5th edn. McGraw-Hill.

Pan B, Kang R, Guo J, Fu H, Du D and Kong J 2019 Fabrication of thin copper substrate by double-sided lapping and chemical mechanical polishing. *Journal of Manufacturing Processes* **44**, 47–54.

Pan J, Yan Q, Xu X, Zhu J, Wu Z and Bai Z 2012 Abrasive particles trajectory analysis and simulation of cluster magnetorheological effect plane polishing. *Physics Procedia* **25**, 176–184.

Sahab ARM, Saad NH, Rashid AA, Noriah Y, Said NM, Zubair AF, Jaffar A, Yusoff N and Hayati SN 2013 Effect of double sided process parameters in lapping silicon wafer. *Applied Mechanics and Materials* **393**, 259–265.

SEMI 3D16 2016 Specification for glass base material for semiconductor packaging (3D16-1116) website. URL http://www.semi.org, https://www.semiviews.org.

SEMI 3D2 2013 Specification for glass carrier wafers for 3ds-ic applications (3D2-0216) website. URL http://www.semi.org, https://www.semiviews.org.

SEMI 3D4 2015 Guide for metrology for measuring thickness, total thickness variation, bow, warp/sori, and flatness of bonded wafer stacks (3D4-0915) website. URL http://www.semi.org, https://www.semiviews.org.

SEMI 3D8 2014 Guide for describing silicon wafers for use as 300 mm carrier wafers in a 3ds-ic temporary bonded-debonded (tbdb) process (3D8-0514) website. URL http://www.semi.org, https://www.semiviews.org.

Shigley JE and Mischke CR 2002 *Mechanical Engineering Design (McGraw-Hill Series in Mechanical Engineering)* 5 edn. McGraw-Hill.

Shimura F 1988 *Semiconductor Silicon Crystal Technology*. Academic Press.

Vandamme R, Xin YB and Pei Z 2000 Method of processing semiconductor wafers. Technical report, SunEdison Semiconductor Ltd. US Patent 6,114,245.

Wang L, Hu Z, Yu Y and Xu X 2018 Evaluation of double-sided planetary grinding using diamond wheels for sapphire substrates. *Crystals*.

Wenski G, Altman T, Winkler W, Heier G and G. H 2002 Doubleside polishing– a technology mandatory for 300 mm wafer manufacturing. *Materials Science in Semiconductor Processing* **5**, 375–380.

Whitehouse DJ 2010 *Handbook of Surface and Nanometrology* 2nd edn. CRC Press.

Wikipedia 2018 Lapping Website. URL https://en.wikipedia.org/wiki/Lapping.

8

Chemical Mechanical Polishing

8.1   Introduction

Chemical mechanical polishing (CMP) is commonly employed in two different sets of polishing occasions. The first one is the CMP process to produce the prime (or premium) wafers from ingots. This CMP process is the last step of wafer manufacturing and wafer surface processing from crystal growth to forming to polishing, as illustrated in Figure 2.1. The prime wafers produced after this CMP step, with inspection and packaging, will be ready for microelectronics fabrication. The other CMP process is sometimes called CMP planarization, referring to the processes of making the surface of wafers flat or planarized at various stages of making microchips in microelectronics fabrication processes. This CMP process is performed to remove excess materials or to create a flat foundation for adding the next layer of circuit features.

In this chapter, the CMP process discussed will be the former one for producing prime wafers.

The CMP process has been employed for years to produce the prime (or premium) wafers from ingots, especially for silicon prime wafers. The CMP process is the last step of wafer manufacturing, after a series of wafer surface processes, to produce prime wafers from "bulk crystal growth" to "wafer forming" and wafer "polishing," as illustrated in Figure 2.1.

A polished silicon wafer with a surface that is particle free with mirror surface finish is required for microelectronics processing to fabricate, using the planar technology, semiconductor devices on the wafer surface. The CMP process utilizes a slurry that consists of chemicals and abrasive micro-grits to process smooth surface of a wafer with material removal involving both chemical etching and mechanical abrasion.

The principles of the CMP process are first presented in this chapter by using a schematic illustration to explain the different components of a CMP machine and their functions. Next, various schemes for polishing are presented and discussed. After that, the polishing pad technology and slurry for polishing are discussed. This is followed by the edge polishing process of wafers.

Wafer Manufacturing: Shaping of Single Crystal Silicon Wafers,
First Edition. Imin Kao and Chunhui Chung.
© 2021 John Wiley & Sons Ltd. Published 2021 by John Wiley & Sons Ltd.

8.2 Chemical Mechanical Polishing (CMP)

Chemical mechanical polishing (CMP), also called chemo-mechanical polishing in some literature, can be viewed as a combination of etching and free abrasive polishing performed on silicon wafers to achieve the necessary quality of wafer surface for semiconductor fabrication. The CMP technology was first presented by the Monsanto company in 1965 [Walsh and Herzog (1965)].[1] Free-abrasive machining of mechanical abrasion alone can cause surface and subsurface damage when inappropriate loads are applied, while chemical wet etching alone cannot attain the required surface finish required for prime wafers and/or planarization for IC fabrication. Chemical reactions with silicon surface can be isotropic or anisotropic; however, most chemicals will etch different crystalline planes with different rates. CMP simultaneously involves both effects of chemical etching and mechanical polishing, with free abrasive machining to produce wafers with the desired surface quality.

After lapping and/or grinding surface processing steps, the wafers with smooth surfaces are polished by using the CMP process (cf. Figure 2.1). The CMP process is performed with a slurry, usually called *colloid*. The polishing process is kept under careful management and control in order to reduce the damage to wafer's crystalline structure by using very low loading on the wafers, especially toward the end of the polishing process. In addition, thermal management is necessary to control the temperature of the polishing environment due to the excess of heat generated from the mechanical friction and chemical reaction.

8.2.1 Schematic Illustration of the CMP Process and System

A photo of CMP equipment in a wafer manufacturing facility (Global Wafers, Co., Ltd.) to produce prime wafers is presented in Figure 2.24, with a zoom-in view on the upper-right corner for the illustration of the detail of the CMP equipment. Figures 8.1 and 8.2 provide schematic drawings of the basic components of CMP to illustrate the operating principles of the process. As illustrated with a front view in Figures 8.1(a) and a top view in 8.2(b) with photo, CMP equipment includes a rotating platen covered by a polishing pad, wafer holders and carriers on top of the polishing pad, retaining rings, pad conditioner and arm, slurry dispenser, and a robot for loading and unloading the wafers (not shown in the figure).

The wafer is mounted and held by the wafer holder and carrier assembly, with the surface to be polished facing the polishing pad. A backing film is in place on the other surface of the wafer, as shown in Figure 8.1(a) with a zoom-in view. Vacuum suction provided by a vacuum chuck is used not only to hold and secure the wafer, but also to keep contaminants in the slurry from reaching the back surface of the wafer

---

1 The concept of applying CMP for wafer planarization in IC fabrication was said to be invented at IBM in the early 1980s by Klaus D. Beyer in an attempt to create a highly planarized silicon wafer surface to facilitate the subsequent lithographic process without much distortion of imaging of masks, by filling the trenches with a dielectric (such as silicon dioxide, SiO_2) and removing the excess material by polishing it off with a CMP operation [Krishnan et al. (2010)]. This is often referred to as CMP planarization in the literature.

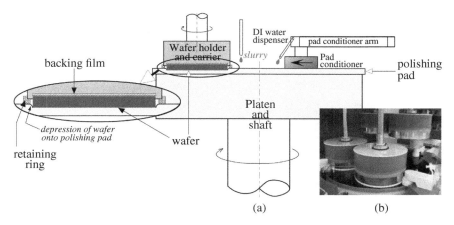

(a)   (b)

Figure 8.1 Schematic of chemical mechanical polishing (CMP) equipment: (a) a front view showing one wafer holder with chuck/carrier, slurry dispenser, DI water dispenser, pad conditioner and arm, platen and shaft, as well as the wafer held by the wafer holder and shaft, with a zoom-in view of the contact interface, and (b) a photo of CMP equipment with wafer holders and chuck/carriers shown in the front view. Source: Imin Kao.

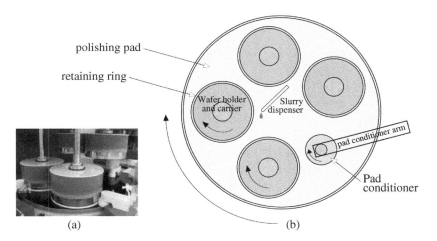

(a)   (b)

Figure 8.2 Schematic of CMP equipment: (a) a photo of CMP equipment with wafer holders and chuck/carriers, and (b) top view showing multiple wafer holders and carriers, retaining rings, polishing pad, pad conditioner, pad conditioner arm, slurry dispenser, and the direction of rotation of each component. Source: Imin Kao.

during the CMP process. A plastic retaining ring, which contains slots for effective slurry flow during polishing, is employed to keep the wafer in place with a horizontal configuration, as shown in the zoom-in view in Figure 8.1(a). The retaining ring must not have contaminants, such as aluminum or copper, in its composition and should not have edges that can scratch the wafer surface.

Both the platen and wafer holder are rotating, typically in the same direction, with a controlled speed for good polishing results. While undergoing the polishing operation, the slurry dispenser on top of the polishing pad continues supplying the polishing slurry that is transported to the polishing site under the wafer holder. At the same time, a pad conditioner oscillates on the polishing pad and acts as a tool for conditioning or dressing the pad and cleaning during the CMP process. An automated process of loading and unloading wafers is performed by a robot. The wafer is kept by the vacuum chuck during loading and unloading. Wafer cleaning is performed after the CMP process.

The silicon wafer, held by the rotating wafer holder and carrier head, presses against the rotating polishing pad, as shown in Figure 8.1(a). As the wafer presses against the polishing pad, it creates a depression in the polishing pad, as illustrated in Figure 8.1(a), while the slots in the retaining ring allow effective flow of slurry to the polishing area of the wafer surface. The slurry reacts with the wafer surface to create a weakened layer that is removed by the mechanical abrasion of the rotating polishing pad. The free abrasive machining process (cf. Section 3.4.2 and Figure 3.6) using the abrasive micro-grits and suspended particles in the slurry provides a gentle process to remove a thin layer of material from the surface of silicon wafer to produce prime wafers with mirror surface finish. The combination of both mechanical action and chemical reaction accomplishes the task of the removal of material from the surface of the silicon wafer. The thickness of the layer of silicon removed in the CMP process can vary, but is typically 10–20 µm. The layer of material removal for CMP in IC fabrication can be in the range of hundreds of nanometers; for example, a layer as thin as 100 nm of copper, metal or other materials.

During the CMP operation, heat is generated from the mechanical abrasion and friction, as well as from the heat of chemical reaction. The polishing speed depends on the temperature of the mechanical and chemical interface. As a result of the heat generation during the polishing process, it is important to manage the temperature. The platen of the CMP equipment has a temperature control system to adjust the temperature between 10 and 70 °C. This is done either by back spraying the platen to cool down the heat generated, or by having a closed-loop circulating cooling system in order to maintain appropriate temperature range for the CMP operation.

The highest quality of surface finish can be obtained by giving a second polish using a very fine (0.05 µm) aluminum oxide suspension, preferably on a self-adhesive cloth pad fixed to a plain stainless steel plate. The CMP process can be analyzed and evaluated by many process variables. They include the quality of polishing pad, slurry contents and flow, pad conditioner, wafer holder and carrier design, vacuum chuck holding the wafer, speed of rotation, temperature control, and the shape of the wafer.

There is a wealth of literature on CMP, including research, patents, and industrial practice. Some are listed here, including Alpsitec (2019); Applied Materials (2020); Bengochea et al. (2018); Dangel et al. (1999); Ensinger (2020); Kaufman et al. (1991); Krishnan et al. (2010); Lai and Lin (2003); Lawing et al. (2015); Qin et al. (2004); Sandhu and Doan (1993); Shiro et al. (2004); Tichy et al. (1999); Wells (1995); Wikipedia (2020); Xin (2003); Zhao and Chang (2002).

8.2.2 Measurement and Evaluation of the Silicon Wafer after Polishing

Several parameters are measured after the CMP process primarily to determine the surface flatness of polished single crystalline wafers. Refer to Section 1.4.4 for details of measuring wafer flatness and a relevant bibliography [SEMI M24 (2012)].

(a) GBIR or the total thickness variation (TTV): GBIR is the same as the TTV, which measures to difference between the maximum and minimum thickness of polished wafers. This measurement traditionally has been an important evaluation of a wafer to assess the maximum variation of the thickness across the entire wafer. Refer to Sections 1.4.1 and 1.4.4, as well as Example 1.4.2 for further information.
(b) Taper: Taper is the difference between the highest and lowest elevation of the polished surface of the semiconductor wafer, using the ideal back surface as the reference. It is usually used as a comparison between the center and edge of the wafer. Taper represents the lack of parallelism between the unpolished surface of the wafer and the selected focal plane on the polished surface.
(c) Measurement methods: The measurement method can be either global, for the entire wafer surface, or site, for the local site. Refer to Section 1.4.4 and Table 1.5 for more details.
(d) Flatness measurements: The flatness measurement can be a total indicated reading (TIR), which represents the *range*, or focal plane deviation (FPD), which represents the *deviation*. Refer to Section 1.4.4 and Table 1.5 for more detail.

The SEMI M24 document [SEMI M24 (2012)] specifies the requirements for virgin polished monocrystalline silicon premium (or prime) wafers with nominal diameter from 150 to 300 mm used for particle counting, metal contamination monitoring, and measuring pattern resolution in the photolithography process in semiconductor manufacturing. Wafers of 300 mm diameter is shipped in accordance with SEMI M45 [SEMI M45 (2017)]. The specification includes the following information:

- Growth method
- Crystal orientation
- Conductivity type
- Dopant
- Wafer surface orientation
- Resistivity
- Diameter
- Fiducial dimensions
- Primary flat/notch orientation
- Edge profile
- Thickness

The SEMI M24 document also provide specifications for polished monocrystalline silicon premium wafers for 180, 130, and 90 nm design rule usage.

Table 8.1 Specifications for polished silicon wafers of 150, 200, and 300 mm.

Wafer specification	150 mm	200 mm	300 mm
Diameter	150 ± 0.2 mm	200 ± 0.5 mm	300 ± 0.2 mm
Thickness	675 ± 15 μm	725 ± 15 μm	775 ± 25 μm
Bow	< 20 μm	< 20 μm	< 50 μm

8.2.3 Specifications for Polished Silicon Wafers

The polished wafers are prepared for CMOS and VLSI/ULSI microelectronic fabrication, as well as MEMS application (especially, the 150 mm wafers). In the following tables, typical specifications for polish wafers are listed.

Table 8.1 lists the basic information of diameter and thickness of single crystalline silicon prime wafers of sizes of 150, 200, and 300 mm.

Table 8.2 lists the specifications, under the "lithography and patterning" requirement, of polished mono-crystalline silicon premium (or prime) wafers for the 180, 130, and 90 nm design rule usage, according to SEMI M24 (2012). In this table, the wafer flatness terminology of SFQR and SFSR are explained in the following. The SFQR flatness is a **s**ite measurement, with the **f**ront surface of the wafer as the reference surface, with the least-s**q**uares reference plane, and with the flatness measurement in **r**ange, or TI**R**. The SFSR flatness is a **s**ite measurement, with the **f**ront surface of the wafer as the reference surface, with the least-**s**quares reference plane on sub**s**ite, and with the flatness measurement in **r**ange, or TI**R**. Refer to SEMI M24 (2012) and Section 1.4.4 for more detail.

Table 8.2 Specifications under the "lithography and patterning" requirement for polished mono-crystalline silicon premium (or prime) wafers for 180, 130, and 90 nm design rule usage, according to SEMI M24 (2012).

Design rule usage	180 nm	130 nm	90 nm
TTV or GBIR	≤ 5 μm	≤ 3 μm	≤ 3 μm
Warp	≤ 75 μm for 200 mm	≤ 75 μm for 200 mm	≤ 75 μm for 200 mm
	≤ 100 μm for 300 mm	≤ 100 μm for 300 mm	≤ 90 μm for 300 mm
Sori	≤ 75 μm for 200 mm	≤ 75 μm for 200 mm	≤ 75 μm for 200 mm
	≤ 100 μm for 300 mm	≤ 100 μm for 300 mm	≤ 90 μm for 300 mm
Site flatness, SFQR*	$\leq$180 nm	–	$\leq$90 nm
SFSR[†]	$\leq$150 nm	$\leq$130 nm	$\leq$65 nm

a) SFQR is measured over a site of 25×25 mm and percentage of usable area, PUA 90% for 180 nm rule, not specified for 130 nm rule, and 25×8 mm and PUA of 95% for 90 nm rule.
b) SFSR is measured over a site of 25×32 mm for 180 nm rule, 25×32 mm for 130 nm rule, and not specified for 90 nm rule.

8.2.4 Types of CMP Processes

The most common CMP equipment is for single-side polishing, as illustrated and discussed in Section 8.2.1. Single-sided polishing using a double-sided polishing mechanism was presented in a patent [Wells (1995)]. In this application, the wafer is polished on both sides concurrently; whereas the surface of other side is protected from being polished by a layer of silicon oxide. Therefore, this double-sided polishing process essentially only produces an outcome of one-sided polishing. In Dangel et al. (1999), the wafer is subjected to double-sided polishing in which the front and back surfaces of the wafer are polished. The wafer is then subjected to single-sided etching in which the back surface, but not the front surface, of the wafer is exposed to a chemical etchant to roughen the back surface of the wafer for handling. Other double-sided polishing through simultaneous or "flip" polishing techniques is employed in industry.

Wafers subject to single-sided polishing are easier to handle because the other side of the wafer is coarse, and is not subject to polishing action. On the other hand, the wafer is more difficult to handle during a simultaneous double-sided polishing process, although it is a very efficient way to reach uniform flatness.

8.2.5 Challenges of CMP Technology

The CMP process has been a very well studied subject in research and technology development over the last two to three decades, especially in the context of wafer manufacturing and microelectronics fabrication. There are many ongoing challenges as the CMP technology evolves. CMP has various challenges including stress cracking, contamination, delaminating at weak interfaces, polishing pad conditioning, and corrosive attacks from slurry chemicals. When CMP is employed for polishing of a localized area in microelectronics fabrication, the challenges include the identification of end points or regions for processing, the amount of material removal, uneven material removal rate due to the characteristics of different materials, and slurry technology. Research and engineering optimization are needed in responding to the challenges.

8.3 Polishing Pad Technology

The polishing pad plays an important role in CMP. The ability to polish more wafers per pad is a critical design consideration to reduce the cost of wafer production. Hence, understanding polishing pad technology becomes an integral part of CMP. Much research has been conducted on polishing pad technology. Only a few are listed here. Polishing pad researchers have studied a number of variables that affect the performance of the polishing pad, such as pressure distribution versus pad deformation, intrinsic polymer properties of pad components, polyurethane and polyester versus pad surface properties, and the conditioning process of the pad versus weariness of the pad [McGrath and Davis (2004)]. Improving polishing pad performance has been reported by optimizing the buffing process of the polishing pad [Park et al. (2006)]. A polishing pad created by reaction injection molding of size controlled gas bubbles within a polyurethane matrix was presented in Preston et al. (2005).

In CMP process, the mechanical properties of the wafer itself must be considered too. If the wafer has a large bow or taper, the pressure will be greater on the edges than that at the center, or vice versa. This can cause a non-uniform polishing action due to the uneven pressure distribution during polishing. Pressure can be applied to the backside of the wafer to press on the entire wafer in order to reduce or remove the bow during polishing.

The polishing pads used in the CMP tool should be rigid enough in order to uniformly press upon the wafer surface during polishing. However, these rigid pads must be kept in alignment with the wafer at all times. Therefore, the actual design of polishing pads is often interlaced layers of soft and hard materials, stacked on top of each other, in order to conform to topography of wafer surface. Typical polishing pads have a roughness of about 50 μm [Wikipedia (2020)]. While in operation, the polish pad makes contact with the wafer surface through its asperities on a randomly rough surface texture.

Polishing pads are mostly proprietary of the manufacturers, and are usually referred to by their trademark names. Generally, these pads are made from porous polymeric materials with a pore size between 30 and 50 μm. The polishing pads are under mechanical abrasion and chemical action in the process; therefore, they must be regularly reconditioned (cf. Section 8.3.1).

8.3.1 Polishing Pad Conditioning

A CMP pad conditioner with a conditioning arm is illustrated in Figures 8.1 and 8.2. The CMP pad conditioner provides the ability to dress the polishing pad by using bonded abrasives, typically a diamond abrasive disc.[2] The pad conditioner is used to maintain the asperity and stability through the removal of worn surface materials, in order to restore the intrinsic structure of the polishing pad and its ability to continuously perform the CMP operation. When the CMP process is in progress, the wafer surface and polishing slurry, with pressure and abrasion, cause erosion to tall asperities into shorter ones and tend to flatten them to the glazed pad surface. Lawing et al. (2015) employed vertical scanning interferometry to measure pad texture, and conduct a study on the influence of pad conditioning on the CMP process. When a pad conditioner is used, the final pad surface is derived from the inherent pad texture (porosity) and the conditioner cutting characteristics (near surface roughness). A good design of pad conditioner and speed of rotation can have a profound impact on the efficacy of the polishing operation provided by a well-conditioned pad in CMP.

---

2 The concept of "dressing" here is similar to the dressing stick discussed in Section 4.4 in Figure 4.5, which is used to dress the diamond bonded abrasive grits on an ID saw.

Patents and research are abound in this topic of pad conditioning. Refer to the following sample references for more reading on pad conditioning Borucki (2002); Chen (1998); Hooper et al. (2002); Lawing et al. (2015); Park et al. (2007); Stein et al. (1996); Tso and Ho (2007).

8.4 Polishing Slurry Technology

Chemical mechanical polishing (CMP) slurries are a mixture containing chemicals and abrasive micro-grits. CMP slurry, also called *colloid*, is composed of colloidal silica (0.125 μm) or aluminum oxide (0.3 μm). The CMP slurry is supplied by slurry dispenser from above the platen continuously onto the polishing pad, as illustrated in Figures 8.1 and 8.2. The highest quality of surface finish can be obtained by giving a second polish using a very fine (0.05 μm) aluminum oxide on a self-adhesive cloth pad affixed to a stainless steel polishing plate. Process parameters relevant to the polishing slurry technology include the size and particle size distribution (PSD) of the abrasive micro-grits and the pH control of the chemicals in slurry. Normally alkalic conditions are used.

The SEMI C96 document provides a standardized test method for measuring the density of CMP slurries at a temperature of $25.0 \pm 0.1\,^{\circ}C$. The SEMI C98 document specifies the measurement of the particle size distribution (PSD) of CMP slurries. Most CMP slurries have a wide range of particle sizes from a few nanometers to several hundreds of nanometers. The certificate of analysis (CoA) for CMP slurries should report the following parameters:

- pH
- % solid content
- Density [cf. SEMI C96 (2018)]
- Viscosity
- Mean particle size
- Manufacturer and model number of the analytical device used to generate data for particle size distribution inference.

Other parameters can also be reported, as follows

- Conductivity
- Trace metals
- Particle size distribution
- Modality of particle size distribution
- Zeta potential.

Relevant references are listed in the following: Allen (1992); Alpsitec (2019); Kamiti et al. (2007); Mullany and Byrne (2003); Pate and Safier (2016); Provder (1997); SEMI C96 (2018); SEMI C98 (2019); Sivanandini et al. (2013).

8.5   Edge Polishing

In Section 9.3, we present in more detail the edge grinding (or edge rounding) process of silicon wafers, with the purpose of rounding the edge of the wafer to prevent flaking or chipping or stress concentration from taking place along the peripheral of the wafer. The edge polishing processes is a process to further polish the wafer edge after the shape has been defined through the edge grinding process, or to repair and/or improve the edge profile during the microelectronics fabrication process. The edge polishing process typically utilizes a grinding wheel with predefined edge profile, abrasive tape with flexibility in contour following, or polishing slurry in free abrasive machining (FAM).

It was reported that defects on the wafer edge can be transferred to the active device area of the wafer during microelectronics fabrication processing. Removal of such defects using edge polishing can potentially increase the overall yield up to 10% [Applied Materials (2008)].

The issues and importance of wafer edge polishing are similar to those of wafer edge grinding, as presented in Section 9.3. Similarly, the standard of "wafer edge profiles" in the SEMI M73 document [SEMI M73 (2018)] for the characteristics of wafer edge profiles in Figure 9.2 should also be followed.

8.5.1   Fundamentals of Edge Polishing

Fundamentally, the edge polishing processes can be classified into two types, according to the nature of abrasive machining (cf. Section 3.4). They can be classified as follows.

1. Bonded abrasive polishing: This types of polishing process utilizes bonded abrasive in various mechanism designs to perform polishing on the rounded edge profile. It can done by a tape with bounded abrasives to rub around the rounded profile of the peripheral edge of silicon wafers of all sizes to perform polishing [for example, Mipox Corp. (2020)]. It can also be a mechanism of small rotating wheels with bonded abrasives operated under compliance loading to polish the rounded edge profiles. Such wafer edge polishing processes can be operated under dry condition without slurry.
2. Free abrasive polishing: This types of polishing process employs free abrasive machining with slurry to perform polishing on the rounded edge profile. Such equipment that uses slurry polishing action to improve wafer edges operates in wet conditions. Thus, it is often equipped with an ability to perform post-polish cleaning to achieve dry-in and dry-out processing (for example, [SpeedFAM (2020)]).

8.5.2   Challenges of Edge Polishing

Various technical challenges in edge polishing include engineering analysis, modeling, understanding the process for continuous improvement, and slurry content and management. Furthermore, edge polishing systems must be able to remove materials, such as metal and dielectric stack residues, while preserving the features of the wafer edge in Figure 9.2, including the apex, shoulder, bevel, and transition to the front and back surfaces of the wafer that may have been compromised by residues.

8.6   Summary

In this chapter, the chemical-mechanical polishing (CMP) process and technology are discussed, with a focus on wafer manufacturing to produce prime wafers. The fundamentals of the CMP process are illustrated with schematic drawings of a CMP machine tool with different views to explain the components and their functions. The measurements and evaluation of a polished monocrystalline silicon wafer are discussed, primarily pertaining to the characteristics of wafer flatness presented in Chapter 1. The specifications of polished silicon wafers according to SEMI standards are presented. Next, the polishing pad technology and polishing slurry technology are discussed with the subject of polishing pad conditioning. Finally, the application of polishing techniques in wafer edge polishing is presented.

References

Allen T 1992 *Particle size analysis* 4th edn. Chapman and Hall, London.

Alpsitec 2019 CMP, chemical mechanical planarization, polishing equipment website. URL https://www.crystec.com/alpovere.htm.

Applied Materials 2008 Applied materials expands into emerging wafer edge cleaning market with innovative inflexion system website. URL http://www.appliedmaterials.com/en-sg/company/news/press-releases/2008/05/applied-materials-expands-into-emerging-wafer-edge-cleaning-market-with-innovative-inflexion-system.

Applied Materials 2020 CMP, semiconductor website. http://www.appliedmaterials.com/semiconductor/products/cmp/info.

Bengochea LV, Sampurno Y, Stuffle C, Han R, Rogers C and Philipossian A 2018 Visualizing slurry flow in chemical mechanical planarization via high-speed videography. *ECS Journal of Solid State Science and Technology* **7**(3), 118–124.

Borucki L 2002 Mathematical modeling of polish rate decay in chemical-mechanical polishing. *Journal of Engineering Mathematics* **43**, 105–114.

Chen LJ 1998 Chemical-mechanical polish (CMP) pad conditioner. US Patent 5,823,854.

Dangel M, Davis E, Fronterhouse D, Smith J, Smith W and Trentman B 1999 Method of improving the flatness of polished semiconductor wafers. WO Patent App. PCT/US1998/025,429 WO1999031723 A1.

Ensinger 2020 Retaining ring website. URL https://www.ensingerplastics.com/en-us/semiconductor/case-study-cmp-retaining-rings.

Hooper BJ, Byrne G and Galligan S 2002 Pad conditioning in chemical mechanical polishing. *Journal of Materials Processing Technology* **123**, 107–113.

Kamiti K, Popadowski S and Remsen E 2007 Advances in the characterization of particle size distributions of abrasive particles used in CMP *Materials Research Society Symposium Proceedings*, vol. **991**.

Kaufman FB, Thompson DB, Broadie RE, Jaso MA, Guthrie WL, Pearson DJ and Small MB 1991 Chemicalmechanical polishing for fabricating patterned w metal features as chip interconnects. *J. Electrochem. Soc.* **138**(11), 3460–3465.

Krishnan M, Nalaskowski JW and Cook LM 2010 Chemical mechanical planarization: Slurry chemistry, materials, and mechanisms. *Chem. Rev* **110**, 178–204.

Lai CL and Lin SH 2003 Electrocoagulation of chemical mechanical polishing (CMP) wastewater from semiconductor fabrication. *Chemical Engineering Journal* **95**, 205–211.

Lawing AS, Chou E and Yamada A 2015 Advanced in CMP pad conditioning *NCCAVS CMPUG Symposium on CMP Technology and Market Trends*, San Francisco, CA.

McGrath J and Davis C 2004 Polishing pad surface characterization in chemical mechanical planarization. *Journal of Materials Processing Technology* **153**, 666–673.

Mipox Corp. 2020 Silicon wafer edge polisher website. URL https://www.pwsj.co.jp/en/semiconductor/mipox.

Mullany B and Byrne G 2003 The effect of slurry viscosity on chemical-mechanical polishing of silicon wafers. *Journal of Materials Processing Technology* **132**, 28–34.

Park J, Kinoshita M, Matsumura S, Park K and Jung H 2006 Study of pad surface treatment to control polishing performance. *SCRIBD*.

Park KH, Kim HJ, Chang M and D. Jeong H 2007 Effects of pad properties on material removal in chemical mechanical polishing. *Journal of Materials Processing Technology* **187-188**, 73–76.

Pate K and Safier P 2016 Advances in Chemical Mechanical Planarization (CMP) vol. 86 Woodhead Publishing Series in Electronic and Optical Materials chapter 12 Chemical metrology methods for CMP quality, pp. 299–325.

Preston S, Hutchins D and Hymes S 2005 Polishing pad and method of making same. US Patent 2005/021854.

Provder T 1997 Challenges in particle size distribution measurement past, present and for the 21st century. *Progress in Organic Coatings* **32**, 143–153.

Qin K, Moudgil B and Park CW 2004 A chemical mechanical polishing model incorporating both the chemical and mechanical effects. *Thin Solid Films* **446**, 277–286.

Sandhu G and Doan T 1993 Method for controlling a semiconductor (CMP) process by measuring a surface temperature and developing a thermal image of the wafer. US Patent 5,196,353.

SEMI C96 2018 Test method for determining density of chemical mechanical polish (CMP) slurries (C96-0618) website. URL http://www.semi.org, https://www.semiviews.org.

SEMI C98 2019 Guide for chemical mechanical planarization (CMP) particle size distribution (PSD) measurement and reporting used in semiconductor manufacturing (C98-1219) website. URL http://www.semi.org, https://www.semiviews.org.

SEMI M24 2012 Specification for polished monocrystalline silicon premium wafers (M24-0612) website. URL http://www.semi.org, https://www.semiviews.org.

SEMI M45 2017 Specification for 300 mm wafer shipping system (M45-1110) website. URL http://www.semi.org, https://www.semiviews.org.

SEMI M73 2018 Test method for extracting relevant characteristics from measured wafer edge profiles (M73-1013) website. URL http://www.semi.org, https://www.semiviews.org.

Shiro K, Hashisaka K and Oka T 2004 Polishing pad. US Patent 6,705,934.

Sivanandini M, Dhami SS and Pabla BS 2013 Chemical mechanical polishing by colloidal silica. *International Journal of Engineering Research and Applications (IJERA)* **3**, 1337–1345.

SpeedFAM 2020 Edge polisher website. URL http://www.speedfam.com/en/products/edge_polisher.html.

Stein D, Hetherington D, Dugger M and Stoud T 1996 Optical interferometry for surface measurements of CMP pads. *Journal of Electronic Materials* **25**, 1623–1627.

Tichy J, Levert JA, Shan L and Danyluk S 1999 Contact mechanics and lubrication hydrodynamics of chemical mechanical polishing. *Journal of The Electrochemical Society* **146**(4), 1523–1528.

Tso PL and Ho SY 2007 Factors influencing the dressing rate of chemical mechanical polishing pad conditioning. *Int J. of Advanced Manufacturing Technology* **33**, 720–72.

Walsh RJ and Herzog AH 1965 Process for polishing semiconductor materials. US Patent 3,170,273.

Wells R 1995 Method for single sided polishing of a semiconductor wafer. US Patent 5,389,579.

Wikipedia 2020 Chemical-mechanical polishing (CMP) website. URL https://en.wikipedia.org/wiki/Chemical-mechanical_polishing.

Xin YB 2003 Modeling of pad-wafer contact pressure distribution in chemical mechanical polishing. Technical report, Wafer Technology, MEMC Electronic Materials, Inc.

Zhao Y and Chang L 2002 A micro-contact and wear model for chemical-mechanical polishing of silicon wafers. *Wear* **252**(3-4), 220–226.

9

Grinding, Edge Grinding, Etching, and Surface Cleaning

9.1 Introduction

In this chapter, several processes of wafer manufacturing are presented. These processes include wafer grinding and edge grinding (or edge rounding) in *wafer forming*, etching in *wafer polishing*, and wafer surface cleaning in *wafer preparation*, as illustrated in Figure 2.1. The grinding process of wafer surface processing is in Section 9.2. The edge grinding process of wafers is presented in Section 9.3. Various topics of etching and cleaning of wafers are presented in Sections 9.4–9.6.

9.2 Wafer Grinding for Surface Processing

As illustrated in Figure 7.6, the wafer grinding process can complement lapping for wafer surface processing, especially for large wafers such as 300 mm silicon wafers. Wafer grinding technique can be broken into rough grinding and fine grinding in wafer surface process. A rough grinding/preliminary planarization can be performed on wafers before they are batched for lapping to reach the required flatness and to remove waviness. (see also discussions in Sections 7.4.2 and 7.5.1).

Because silicon material is fragile and brittle after shaping into a wafer, one of the main challenges of wafer grinding is meeting the required high capacity to obtain economical yields, and be able to compete with the current batch lapping technology. In addition, the finished wafer product must have precise thickness in the range of microns that is suitable for micro-fabrication applications. In the case of semiconductor applications, the machined surface of the wafer must exhibit minimum roughness and waviness in order to preserve the desired electrical circuits and characteristics during IC fabrication.

9.2.1 Wafer Grinding Methods

Wafer grinding techniques can be categorized by their processes and methods, as will be discussed in the following sections. A list of relevant and useful reference and patents include Brandt (1987); Cong et al. (2009); Dobrescu et al. (2011, 2012); Gao et al. (2010); Li et al. (2006); Liu et al. (2002, 2007); Lu et al. (2007); Nishiguchi

Wafer Manufacturing: Shaping of Single Crystal Silicon Wafers,
First Edition. Imin Kao and Chunhui Chung.
© 2021 John Wiley & Sons Ltd. Published 2021 by John Wiley & Sons Ltd.

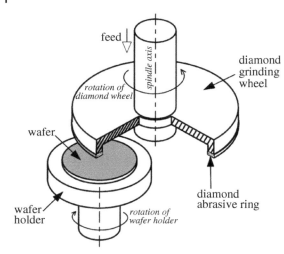

Figure 9.1 Schematic of a wafer on a wafer holder and a diamond grinding wheel of a single surface grinding machine, for the illustration of the wafer grinding process.

et al. (1991); Pei and Fisher (2001); Pei et al. (2008); Sun et al. (2004); Tabuchi (1984); Vandamme et al. (2000); Wang et al. (2018); Young et al. (2006). There are companies that produce wafer surface grinding machines, such as Koyo Machine Industries (2019); SVM (2018); Wolters (2010); Yamazaki (2006) and others.

A schematic of a wafer surface grinding machine and process is illustrated in Figure 9.1. The schematic illustrates a wafer on a wafer holder which may be stationary or rotating at a prescribed speed. A diamond grinding wheel which typically has a ring of diamond abrasives, as shown in Figure 9.1, rotates and down-feeds towards the wafer surface to remove a layer of material from the surface of the mounted silicon wafer on the wafer holder. The diamond grinding wheel and the wafer holder typically rotates in opposite directions, as shown. Such surface grinding is also applied in back grinding of the back surface of wafers with the completed IC fabricated to thin the wafers before dicing. Based on the simplified illustration in Figure 9.1, various permutations of the designs and configurations are discussed in the following:

- Wafer holder mechanism: The wafer holder mechanism can hold a single wafer or multiple wafers, with different arrangements. In addition, the wafer holder can be stationary or rotating at a prescribed speed and direction. The wafer holder with multiple wafers can have revolute or prismatic movement with respect to the grinding wheel.
- Grinding wheel: The grinding wheel can be a single Blanchard-type grinding wheel with a diamond abrasive ring at the edge, as illustrated in Figure 9.1. The grinding wheel can also contain two concentric rings with the inner ring for rough grinding, composed of more coarse diamond abrasive grits, and the outer edge ring for fine grinding, composed of finer diamond abrasive grits. Multiple grinding wheels can also be employed in a simultaneous grinding operation (like the batched lapping operation) to speed up the production.
- Motion of the wafer holder and grinding wheel: The directions of rotation of the wafer holder and grinding wheel are typically opposite to each other. Figure 9.1

illustrates the revolute motion of the wafer holder, with rotary grinding wheel(s). However, the prismatic motion of the wafer holder can also be realized. In most designs, the grinding wheel is the one that provide machining "feed" onto the surface of wafer to perform the material removal.

Single Surface and Double Surface Grinding Methods

The single surface grinding process is schematically illustrated in Figure 9.1. A good review of the three types of single surface grinding machines is provided in Pei et al. (2008). The three types that were developed for back grinding for thinning wafers before dicing, and later for the surface grinding of as-sliced wafers, are:

1. Blanchard type
2. Creep-feed type
3. In-feed type.

The single surface grinding process is an extension to grinding of the as-sliced wafer surface. Such methods were also extended to enable double-sided grinding.

Batched Surface Grinding Methods

When the wafer holder accommodates multiple wafers for batch processing using a rotating grinding wheel, the configuration is batched surface grinding. While the grinding wheel rotates with respect to its spindle axis and advances to the wafer surface (or call "feed"), the plurality of wafers on the wafer holder can be configured with two different types [Brandt (1987)]:

1. Revolute type of batch wafer holder: In this configuration, the wafer holder with multiple wafers rotates with respect to the grinding wheel. The feed can be accomplished by the movement of the rotating grinding wheel, or by the movement of the wafer holder with respect to the grinding wheel for material removal. Such distance of feed (or call "infeed") in the direction of the axis of rotation is controlled or pre-determined for a successful grinding operation of wafer surface processing.
2. Prismatic type of batch wafer holder: In this configuration, the wafer holder with multiple wafers moves prismatically, typically in a linear reciprocating motion. The feed can be accomplished by the movement of the rotating grinding wheel with respect to the reciprocating wafer holder for material removal. The grinding wheel is fed onto the wafer holder by controlled or pre-determined distance (or "infeed") with each stroke of the prismatic wafer holder to remove a shallow layer of material from wafers by the peripheral edge of the grinding wheel.

In both configurations, multiple grinding wheels can be employed to process many wafers on the wafer holder at the same time. Although the batched surface grinding methods provide improvement in yields of wafer surface processing, as compared with the single side grinding method, they have been proven to be slow, with low yields compared with conventional batch lapping, as some studies show. Nevertheless, the batch grinding methods can produce wafers with relatively good surface quality.

Continuous Surface Grinding Method

The continuous surface grinding method is an extension of the batched surface grinding method in which the wafers on the wafer holder continuously and sequentially pass through multiple stages of grinding wheels that are positioned at successive prescribed heights (providing the infeed) to continuously remove layers of materials from the surface of the wafers on the wafer holder. Unlike the batched surface grinding method, each wafer undergoes only a single pass during the continuous surface grinding. The continuous surface grinding method can be configured with wafer holders moving in a revolute or prismatic mechanism.

Although the rate of material removal is relatively higher in the continuous surface grinding methods than that in the batched surface grinding method, the quality of the wafers produced is generally considered to be no better than the quality of the batch surface grinding, according to some studies.

Disk Grinding Method

In contrast to the schematic in Figure 9.1 and the batched or continuous surface grinding methods, the disk grinding method or cross grinding method were designed to utilize the entire grinding disk surface or increase the radial width of the the ring of grinding wheel to cover the entire surface of the wafers with the planar face of the diamond grinding surface.

Unlike surface grinding wherein the wafer holder undergoes an infeed movement towards the grinding wheel during the grinding operation, the wafer holder remains stationary and fixed with respect to the axis of rotation of the grinding wheel during cross grinding or disk grinding. It was claimed that the cross grinding or disk grinding process on thin wafers can produce better results than the batched methods [Brandt (1987)].

9.2.2 Grinding Wheel Technology

The major considerations of grinding wheels for the grinding processes of silicon wafers include:

- The appropriate diamond abrasive grain size for coarse or fine grinding.
- The grinding force should be low and constant.
- The grinding wheel should have a reasonable service life.
- The surface of the grinding wheel must be flat, and may need to be *trued* regularly.
- Coolant is required during the grinding operation. De-ionized water is conventionally used as a coolant for grinding.
- The outcomes of grinding (i.e. the ground wafers) should have very good flatness, with the surface of ground wafers having submicron total thickness variation (TTV).
- The surface and subsurface damage of ground wafers should be minimized.

A key requirement of wafer manufacturing is to produce extremely flat surfaces on wafers of size up to 350 mm in diameter. A TTV of less than 15 μm is strictly demanded by the industry for a 0.18 μm feature size in the IC process. Furthermore, the surfaces should be smooth with a roughness of $R_a < 10$ nm, and have minimum subsurface damage before the final etching and polishing processes. The end product should have crack-free mirror surfaces with a micro-roughness less than 1.8 Å [Young et al. (2006)].

Experimental results showed that the grain size of the diamond abrasive grits has large influence on the subsurface damage of silicon wafers. In addition, the ground wafers processed without spark-out during grinding machining have shorter radial or lateral subsurface cracks, depending on crystalline orientation, than the ground wafers with spark-out during machining. Nevertheless, the depth of subsurface damage in different crystal orientations and radial direction becomes more uniform with spark-out during grinding machining [Gao et al. (2010)].

9.2.3   Types of Grinding Operations

In addition to the different methods of wafer grinding presented in Section 9.2.1, different types of grinding operations are discussed in this section. There are many patents that proposed various embodiment and configuration of designs.

Single-sided Grinding
The single-sided grinding method was presented in Section 9.2.1 with schematic illustration in Figure 9.1. There are many variation of designs of grinding mechanism for single-sided grinding, however.

Simultaneous Double-sided Grinding
The simultaneous double-sided grinding (SDSG) type was introduced in 1990s [Li et al. (2006)]. Another type of simultaneous double-sided grinding is called double-disk grinding (DDG), which utilizes the entire disk surface for grinding operation. In both simultaneous double-sided grinding types (SDSG or DDG), wafers are simultaneously processed on both sides, with the wafer floating freely between the two grinding wheels or disks mounted on collinear spindles on the opposite sides of the wafer. The grinding wheels or disks are guided axially along the axis of rotation, by either water cushion (hydrostatic) or air cushion (aerostatic), with nearly no axial force on both sides of the wafer surface. They are prevented from floating away radially by a surrounding thin guide ring or radial spokes. Both vertical and horizontal spindles can be employed in combination with special diamond grinding tools. SDSG or DDG types have the challenge of sustainable mechanisms to hold wafers, as opposed to the single-sided grinding type, to produce consistent outcomes.

A type of simultaneous double-sided grinding using the lapping kinematics of planetary pad grinding (PPG) was design and produced [Wolters (2010)]. In this design, bonded abrasive grinding pads are used to simultaneously process both sides of wafer with good GBIR[1] (down to 500 nm) and SFQR (down to 100 nm), exceeding the requirements for the 22 nm fabrication technology.

---

1 Refer to Section 1.4.5 for the definitions of GBIR and SFQR in nanotopology.

Fine Grinding

Fine grinding utilizes very fine diamond abrasive grits (cf. Section 3.5) to achieve fine surface finish of wafers. Fine grinding is sometimes integrated with the same grinding process with coarse grinding. For example, the schematic of the single-side, single-wafer grinding illustrated in Figure 9.1 can have an additional coarse diamond abrasive ring between the outer fine diamond abrasive ring (shown in the figure) and the spindle axis. The grinding process will first place the inner coarse grinding ring on the wafer surface to perform coarse grinding to remove material from the wafer surface quickly, followed by a slower infeed using the fine grinding ring to achieve the prescribe surface quality.

9.2.4 Technical Challenges and Advances in Grinding

Because silicon material is fragile and brittle after being shaped into a large wafer, technical challenges in grinding include the modeling and understanding of the grinding processes, the design of the machine tool and engineering, the grinding wheels, the wafer support systems, and the grinding wheel technology.

One main challenge of the wafer grinding process is to have high enough throughput to obtain economical yields, and be able to compete with the current batch lapping technology. In addition, the finished wafer product after the grinding process must have precise thickness in the range of microns that is suitable for micro-fabrication applications. In the case of semiconductor application, the machined surface of wafer must exhibit minimum roughness and waviness in order to preserve the desired electrical circuits and characteristics.

Various grinding methods have been found to produce finished wafers in which the depth of damage can be greater than desired. Moreover, constant and repeated criss-cross patterns of tool marks resulting from the grinding machining operations are often visually evident on the surfaces of the processed wafers. These tool marks significantly increase the possibility of the wafer rupturing during handling and use. Moreover, when the wafers are used as substrates or carriers of electronic circuits, such tool marks can undesirably affect the electrical characteristics of the wafers. The subsurface damage in the region of the ground wafer surfaces and the visually evident tool marks also have a significant adverse affects on the subsequent treatment and diffusion operations to which the wafers are subjected. The less the depth of subsurface damage to wafer, the thinner the wafer can be ground using grinding for surface processing.

9.3 Edge Grinding

Silicon is a brittle material and is prone to various issues of brittle materials, such as fracture, chipping edge, and crack propagation under tensile stress. Such mechanical properties make it easy for silicon wafers to have flaking along the peripheral edge. Two major issues will occur in down-stream processing if the edge treatment is not performed properly [Axus Technology (2013)]. The flaking edge of wafer can cause a local effect to the wafer itself and a global chain reaction to the nearby wafers in the processing and micro-fabrication environment.

- Effect on the wafer itself: Micro-cracks and flaking at the edge of the wafer cause stress concentration, and will facilitate the growth and propagation of cracks to the rest of the wafer material, especially when tensile stress is applied to the wafer during handling and/or processing. This often sudden crack propagation, typically under tensile stress, will result in the individual wafer being lost, broken, or otherwise destroyed beyond usability. As a consequence, a broken wafer with crack lines will also cause the loss of fabricated IC on the wafer, as well as the subsequent cleanup required for the fab environment.
- Effect on the global wafer environment: When wafer edge flaking or edge chipping is generated in the global environment of IC fabrication, it will first cause contaminants in the pristine environment within the clean room and process chamber. When the flakes and cracks of individual wafer worsen to cause the gross breakage of the wafer, it will have disastrous consequences on other wafers in the process chamber and work area of the fab.
- Defects on the wafer edge can be transferred to the active device area of the wafer during microelectronics fabrication processing. Removal of such defects using edge polishing can potentially increase the overall yield up to 10% (cf. Section 8.5).

It is therefore necessary to prevent edge flaking or chipping from occurring on the wafer peripheral edge to ensure its shape integrity for handling and processing. Edge grinding is the step in wafer manufacturing to polish the peripheral edge of wafer after slicing. As illustrated in Figure 2.1, edge grinding (or edge rounding) is typically performed after the wafer slicing process in order to protect wafers from crack propagation and breakage in the subsequent wafer manufacturing processes.

9.3.1 Fundamentals of Edge Grinding

Edge grinding is sometimes referred to as edge rounding, with the purpose of rounding the edge of the wafer to prevent flaking or chipping or stress concentration from taking place on the peripheral of the wafer, as discussed above. The main purpose of wafer edge profile designs is to keep the wafer from chipping. Many variations of wafer edge profiles have been practiced in the wafer manufacturing industry, such as a round or bullet edge profile [cf. Axus Technology (2013); Chapman Instrument (2013)].

SEMI (semi.org) defines the standard of "wafer edge profiles" in the document SEMI M73-1013 [SEMI M73 (1986)] for the characteristics of wafer edge profiles. Figure 9.2 is an illustration adapted from the specification of [SEMI M73 (1986)] on wafer edge profiles. The wafer edge is identified by the front wafer surface line (on the top of the figure) and back wafer surface line (at the bottom of the figure), separated in the middle by the centerline, or the reference line. The origin is at point O, shown in Figure 9.2, at the intersection of the centerline and the apex. The apex is shown in the figure as the vertical line at the peripheral edge of the wafer. Point A is the intersection of the front wafer surface line and the front bevel line, as shown in Figure 9.2. Point B is the intersection of the front bevel line and the front shoulder formed by the arc between points B and C, having a radius of r, centered at point M. Point C, at the upper end of the straight vertical line of the apex, is also at the intersection of the apex line and the front shoulder, as shown.

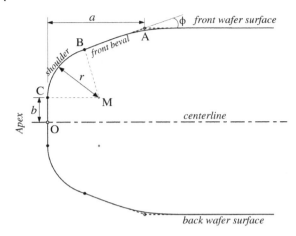

Figure 9.2 Schematic of the wafer edge profiles and its relevant characteristics, as defined by semi.org.

Note that the same terminology is also defined for the bottom half of the wafer edge profiles. The profile at point A is rounded, as shown in Figure 9.2, instead of a sharp edge shown by the dashed lines in the figure. Point A is the point of intersection above the rounded edge. The length a is identified as the "edge width" measuring the total length of the edge profile; the length b is the "apex length"; and r is the shoulder radius.

Based on the basic profile in Figure 9.2, three edge profile types are defined to identify the main types of edge profiles.

1. Type A has a long edge width a, a small shoulder radius, r, and a long apex length b.
2. Type B has a short edge width a, a small shoulder radius, r, and a long apex length b.
3. Type C has an intermediate edge width a, a large shoulder radius, r, and a short apex length b or almost no apex length.

The value ranges of the parameters of the 300 mm wafer are tabulated in Table 9.1. Only one set of value ranges is defined for the 450 mm silicon wafers, as stipulated in the documents SEMI M73 and M1 [SEMI M1 (2017); SEMI M73 (1986)]. The value ranges for 450 mm wafers are listed in Table 9.2.

Tables 9.1 and 9.2 outline the range of values of the parameters of wafer edge profiles, a, b and r. However, each parameter may assume different values within the value ranges in each type, including the types A, B, and C for 300 mm or smaller

Table 9.1 The wafer edge profiles of 300 mm wafers, with value ranges of a, b, and r associated with types A, B, and C

Types	Edge width, a	Shoulder radius, r	Apex length, b
Type A	300–400 μm	200–260 μm	100–175 μm
Type B	200–270 μm	200–260 μm	100–175 μm
Type C	240–310 μm	300 to $t/2^*$	0–100 μm

*The length $t/2$ in the shoulder radius r of type C represents half the thickness of the wafer.

Table 9.2 The wafer edge profiles of 450 mm wafers, with value ranges of the parameters a, b, and r

	Edge width, a	Shoulder radius, r	Apex length, b
For 450 mm wafers	300–400 μm	160–245 mum	120–260 mum

Note only one type of edge profile is defined according to the specifications for 450 mm silicon wafer in SEMI M1.

wafers, and the one type for 450 mm wafers. This encompasses a variety of variations of the actual wafer edge profiles employed in the wafer manufacturing industry. Consult SEMI M73 (1986) for more details.

9.3.2 Technical Challenges in Edge Grinding

The edge grinding process is typically performed using a profiled diamond grinding wheel. There are several challenges associated with the edge grinding process identified by industry (*cf.* [Axus Technology (2013)]). These technical challenges include the following:

- The grinding wheel technology: This includes the choice of the diamond abrasive grits, bonding matrix, and concentration of the composition used to make the diamond edge grinding wheel. In addition, the profiled edge grinding tool must conform to the standards of wafer edge profiles and be able to maintain its shape and profile.
- The machining process: This refers to the challenges in the actual machining process and post-process inspection of the grinding wheel and its operations. The edge grinding tool may need to be *trued* regularly to maintain efficiency of machining. It includes uneven grinding, uneven wear of the grinding wheel with a distorted profile, edge flaking or chipping during grinding, and others.
- Compliance to the SEMI standards: This includes the use of improper angle of the edge profile as compared with the profiles of SEMI M73, profile lengths, radius of curvature, and process control to hold tolerances to SEMI specifications.

In addition to these challenges, the edge grinding process is a time-consuming machining process and will affect the total time needed to produce prime wafers in wafer manufacturing. Management of quality and consistency of the edge contouring is important for efficient process control.

9.4 Etching

Prior to chemical etching, silicon wafers typically exhibit surface and subsurface defects such as embedded particles, contaminants, defects in diffusion, fractures, stack faults, stress-induced crystalline imperfections, and/or physical damage. The physical damage, which includes micro-cracks, fractures, or stress-induced crystalline imperfections, may be induced by stress applied to the wafers during wafer forming, including processes from crystal growth to the surface planarization processes such as CZ growth, slicing, lapping, grinding, and edge rounding or profiling, as illustrated in Figure 2.1. These defects generally occur on the surface of wafers to about 2.5 μm beneath the surface. In order to remove these defects, a layer of material with a depth of 2.5 μm is typically removed from the surface of the wafers using an acidic and/or caustic chemical etchant. Such etching process can remove the undesirable embedded particles, contaminants, defects in diffusion, and physical damage contained in the near-surface layer of silicon wafer. After etching, silicon wafers will undergo polishing and cleaning processes.

9.4.1 Acid Etching

Modern etching technology for silicon utilizes the chemistry of formation and dissolution of silicon dioxide (SiO_2); therefore, chemical etchants in silicon surface processing include hydrofluoric acid (HF), as shown in Equations (9.1) and (9.2). With acid etching, a silicon wafer is dipped in an acidic mixture of nitric acid (HNO_3) and HF and a diluent ether water or acetic acid in an acid etching process. The chemical reaction is in the following formula [Kulkarni and Erk (2000)]:

$$Si + 4\ HNO_3 \rightarrow SiO_2 + 4\ NO_2 + 2\ H_2O \tag{9.1}$$

$$SiO_2 + 6\ HF \rightarrow H_2SiF_6 + 2\ H_2O. \tag{9.2}$$

Oxidants other than the nitric acid can also be used in acid etching.

9.4.2 Caustic Etching

Caustic etching offers certain advantages over acidic etching because of the commercially available caustic chemicals. Commercially available sources include sodium hydroxide (NaOH), potassium hydroxide (KOH), and similar hydroxide etchants. However, these commercial sources typically contain a significant concentration of nickel, copper, and other metals, which are considered as contaminants for silicon processing. These contaminants can be diffused into silicon wafers during the etching process, causing undesirable side effects.

Various approaches are used to reduce the potential for contamination. The caustic etching solution can be pretreated with a chelating resin to reduce the metal ion concentration in the etchant before the etchant contacts the wafers in an etchant tank or equivalent apparatus [Nakano et al. (2000)]. However, this approach introduces an additional processing step and fails to address the problem of metallic impurities introduced into the caustic etching solution after the pretreatment during the etching process [Stinson et al. (2011)]. A caustic solution of alkali metal solution with sodium hydroxide (NaOH) or potassium hydroxide (KOH) can be used [Moldovan and Ilie (1996)].

9.4.3 Preferential Etching

Structural defects can be formed in silicon wafer during its bulk crystal growth, wafer forming, wafer polishing, and/or wafer preparing processes (cf. Figure 2.1) or induced by electronic device processing. These defects can affect the performance of the circuitry fabricated on the silicon wafer. Defects can take the form of point defects, dislocations, slip, planar defects (stacking faults), shallow pits, precipitates, or processing anomalies (for example, areas of different dopant type). Material defects may indicate general quality issues and may not be directly related to a specific failure. However, dopant issues are generally a problem if they are found anywhere on the active area of the IC. Etching often can be used to evaluate wafer quality or initiate failure analysis of an errant device structure [SEMI MF1809 (2010)]. Analytical methods based on preferential etching are frequently used for evaluating such defects in silicon wafers and for assessing the quality of the IC active layers. Preferential etching technique uses the HF for the dissolution of SiO_2, as indicated by Equation (9.2). The dissolution process is enhanced at and around defects due to weakened bonds and coalescence of impurities around them. As a case in point, problems can often be detected in processing memory devices in IC fabrication. For example, in a planar DRAM memory cell, communication problems between adjacent cells can be caused by stacking faults lying on the field oxide isolation between storage capacitors [Semitracks, Inc. (2020)].

The most common etching solutions are Dash, MEMC, Schimmel, Secco, Sirtl, Sopori, Wright, and Yang. The readers may refer to the following reference for more information Chandler (1990); Dash (1956); Hull (1999); Schimmel (1979); Schimmel and Elkind (1978); Secco d'Arragona (1972); Sirtl and Adler (1961); Sopori (1984); Wright-Jenkins (1977); Yang (1984); Válek and Šik (2012)] and the following SEMI document [SEMI MF1725 (2010); SEMI MF1726 (2010); SEMI MF1727 (2010); SEMI MF1810 (2010); SEMI MF1809 (2010). A few common etching solutions are discussed in the following, with a temperature maintained at 20°C during etching process. The chemicals shall have the following normal assay [SEMI MF1809 (2010)]:

Acetic acid	Chromium trioxide	Copper nitrate	Hydrofluoric acid	Nitric acid
>99.7%	>98%	>98%	$49 \pm 0.25\%$	70–71%

Acetic acid = CH_3COOH; chromium trioxide = CrO_3; copper nitrate = $Cu(NO_3)_2$; hydrofluoric acid = HF; nitric acid=HNO_3.

Dash Etch and Modified Dash Etch

The Dash etch is a chromium-free etching solution, primarily used for junction delineation [Dash (1956)], with target surfaces of (100) and (111). The junctions are stained according to their doping concentration when held under a strong light source for a few tens of seconds. It is specified based on a standard solution with a recipe of

$$HF:HNO_3:acetic \ with \ 1:3:10.$$

The Dash etch does not accentuate polishing damage. The approximate etch rate is 0.13 μm min^{-1} on a (100) surface and 0.005 μm min^{-1} on a (111) surface [Hull (1999)].

The modified Dash etch is a chromium-free etching solution. It is specified based on a standard solution with a recipe of

$$HF:HNO_3:acetic:H_2O \ with \ 1:3:12:0.17 \ and \ AgNO_3 \ 0.005 \ to \ 0.05 \ g \ L^{-1}.$$

The approximate etch rate is 1 μm min^{-1} [SEMI MF1809 (2010)].

MEMC Etch

The MEMC etch, also identified as "copper-3" solution, is a chromium-free etching solution [SEMI MF1809 (2010)]. It is specified based on a standard solution with a recipe of

$$HF:HNO_3:acetic:H_2O:Cu(NO_3)_2 \cdot 3H_2O \ with \ 36:25:18:21:$$
$$1 \ g \ per \ 100 \ mL \ of \ total \ volume$$

The etch rate is 1 μm min^{-1} with agitation and 5 μm min^{-1} without agitation.

Schimmel Etch and Modified Schimmel Etch

The Schimmel etch is a chromium-containing etching solution, with a target surface of (100), for a resistivity range of 0.6–1.5 Ω cm [Schimmel (1979); Schimmel and Elkind (1978)]. The Schimmel etch is good for delineating dislocations, slip planes, swirl defects, stacking faults, and oxygen clusters in (100) silicon. This etch is essentially a Sirtl etch (see below) that has been buffered with acetic acid. Delineation of the doped regions occurs in about 5–15 min, with the more lightly doped regions appearing first. It is specified based on a standard solution with a recipe of

$$HF:CrO_3(5 \ M) \ with \ 2:1.$$

The approximate etch rate is 1.8 μm min^{-1} [Hull (1999)]. The modified Schimmel etch, for a resistivity below 0.6 Ω cm, is specified based on a standard solution with a recipe of

$$HF:CrO_3 \ (1 \ M):H_2O \ with \ 2:1:1.5.$$

The approximate etch rate is 1.8 μm min^{-1}.

Secco Etch

The Secco etch is a chromium-containing etching solution, with a target surface of (100) [Secco d'Arragona (1972)]. It is specified based on a standard solution with a recipe of

$$HF:K_2 \ Cr_2 \ O_7 \ (0.15 \ M) \ with \ 1:1.$$

The approximate etch rate is 1 μm min^{-1}.

Sirtl Etch

The Sirtl etch has a target surface of (111) [Sirtl and Adler (1961)]. It is specified based on a standard solution with a recipe of

$$HF:CrO_3 \ (5 \ M) \ with \ 1:1$$

The approximate etch rate is 3 μm min^{-1}.

Sopori Etch

The Sopori etch is primarily used in polycrystalline silicon [Sopori (1984)]. It is specified based on a standard solution with a recipe of

$$HF:HNO_3:acetic \; with \; 36:1-2:20$$

The approximate etch rate is 5–20 $\mu m \; min^{-1}$.

Wright Etch

The Wright etch is a chromium-containing etching solution, with target surfaces of (100) and (111) [Wright-Jenkins (1977)]. The Wright etch can delineate oxidation-induced stacking faults, dislocations, swirls and striations with minimum surface roughness or extraneous pitting. It has been demonstrated that the Wright etch is superior in revealing stacking faults and dislocation etch figures when compared with those revealed by Sirtl and Secco etchings. It is specified based on a standard solution with a recipe of

$$HF:HNO_3:CrO_3 \; (5 \; M):Acetic:H_2O:Cu(NO_3)_2 \cdot 3H_2O \; with \; 2:1:1:$$
$$2 \; g \; per \; 240 \; mL \; of \; total \; volume.$$

The best results are obtained by first dissolving the copper nitrate in the given amount of water; otherwise, the order of mixing is not critical. The approximate etch rate is 0.6–1 $\mu m \; min^{-1}$.

Yang Etch

The Yang etch [Yang (1984)] is used with target surfaces of (100), (110), and (111). The Yang etch can delineate all common defects. It is specified based on a standard solution with a recipe of

$$HF:CrO_3 \; (1.5 \; M) \; with \; 1:1.$$

The etch rate is 1.5 $\mu m \; min^{-1}$ on the (100) and (111) surfaces (not specified on the (110) surface) with no need for agitation.

9.4.4 Technical Challenges and Advances in Etching

Due to the nature of the chemical reactions, different outcomes of surface can result from acid etching and caustic etching. The acid etching process is isotropic and diffusion controlled, and generally yield relatively smooth surface that are free of faceted features. However, acid etching can produce a relatively poor uniformity of silicon surface, which results in significant total thickness variation (TTV). On the other hand, the caustic etching process is anisotropic and reaction controlled. The caustic etching often produces a good uniformity of silicon removal, with a low TTV; however, it is characterized by by a surface topography generated, with faceted surfaces, and higher roughness as a result. The roughness in caustic etching results from the characteristic features (or facets) formed on a surface, such as squares (or pillows) for a (100) wafer and triangles for a (111) wafer. The dimension of these features can increase with the the time of etching.

It is desirable to obtain silicon wafers with

- A good uniformity across the wafer surface (low TTV)
- A low surface roughness which is free of surface features.

Various research, schemes and modeling are conducted to this end to understand and improve the etching processes. In addition, the chemicals used in etching processes are very harmful to the environment. It is, therefore, important to find alternatives and/or reduce the environmental impacts of the etching processes.

9.5 Surface Cleaning

Fabrication of semiconductor microelectronic devices depends on a very clean wafer substrate surface. Any impurity or contamination must be avoided to increase device performance in semiconductor application. It is very important that wafers are cleaned before high-temperature processing involved in oxidation, diffusion and chemical vapor deposition (CVD) of silicon wafer fabrication. Therefore, a critical step before polished wafers are ready for packaging is to clean wafer surfaces so as to be suitable for device fabrication. The cleaning process has to be simple and yet environmentally sustainable, and be able to remove organic and inorganic impurity and surface contaminants in order to meet the requirements of device fabrication.

An excellent reference of wafer cleaning technology was provided by Kern (1990) who invented the standard wafer cleaning method that has been followed by the industry since it was developed at RCA and introduced to device fabrication in 1965, and published in 1970 [Kern and Puotinen (1970)]. The RCA wafer cleaning will be presented in Section 9.6.

9.5.1 Impurities on the Surface of a Silicon Wafer

There are two types of impurities on silicon wafers, enumerated as follows [Kern (1990)].

1. Contaminant films and discrete particles, which can be classified as:
   (a) Molecular compound: mostly particles and films of condensed organic vapors from lubricants, greases, photoresists, solvent residues, components from plastic storage containers and metal oxides or hydroxides.
   (b) Ionic materials: cations and anions, such as sodium ion, fluoride ion and chlorine ions from inorganic compounds that may be physically absorbed or chemically bonded on the wafer surface.
   (c) Metal contaminants: various metal species contaminants such as gold, copper and chromium that may be contained in etch solutions, chemicals, or debris from equipment or process.
2. Absorbed gases that are of little practical consequence in wafer processing

The consequences of wafer impurities in semiconductor application include the following:

- Results in poor adhesion of deposited layers in epitaxial deposit by molecular contaminant films on the wafer surface.
- Produces twining dislocations, stacking faults and crystal defects in epitaxial growth by high concentration ions impurity.
- Results in electrical defects, device degradation and yield loss with ionic and metal impurity.
- Causes blocking in photolithography and etching or rinsing due to particle impurity.
- Leads to pinholes, material voids, cracks by particles that are present during film growth or deposition.

Refer to the following for more information: Burkman et al. (1988); Eisele and Klausmann (1984); Kern (1970, 1990); Kern and Deckert (1978); Kern and Puotinen (1970); Kern and Schnable (1987); Khilnani (1988); Licciardello et al. (1986); Monkowski (1987); Ruzyllo (2010); Schmidt and Pierce (1981); Slusser and MacDowell (1987); Stacy et al. (1981).

9.5.2 Various Cleaning Steps in Wafer Process Flow

Various cleaning techniques and steps in wafer process flow are discussed in the following.

Cleaning Techniques for Large Particles

Mechanical methods are typically employed as the cleaning techniques for removing large particles, especially those particles produced during the wafer forming processes of slicing, lapping and grinding (cf. Figure 2.1). Distilled water is best for removing polymeric particles and one-to-one mixture of ethanol acetone is best for the removal of inorganic particles [Menon et al. (1989)].

Brush Scrubbing

The principle of this cleaning is to dislodge particles using hydrodynamic means with brushes made of hydrophilic material, such as nylon [Burggraaf (1981)]. This can be performed by wafer scrubbing machines. Distilled water or isopropyl alcohol is applied to the wafer to help wash off the particles. It is important to maintain a thin layer of fluid between the wafer and brush in order to prevent the surface from scratching [Shwartzman et al. (1985)].

High Pressure Fluid Jet Cleaning

This cleaning method utilizes a high-speed jet of distilled water or organic solvent to sweep over the wafer surface to shear off particle smaller than micron size. The pressure of the high-speed jet of fluid can be as high as 400 psi [Bardina (1988); Burggraaf (1981)]. This method is effective in removing particles and penetrating into dense topography; however, the wafer can be damaged with improper pressure and operation [Skidmore (1987)].

Utrasonic Wash

The ultrasonic wash uses sonic energy of 20 kHz and above to remove particles [Bardina (1988)]. The ultrasonic wash can be applied using an ultrasonic bath with silicon wafer immersed. It was reported that DI water was best for removing polymeric particles, while a 1:1 mixing ratio of ethanol-acetone was better for inorganic particles [Menon et al. (1989)].

Choline Cleaning

Unlike mechanical means of cleaning discussed above, choline cleaning is a chemical treatment that removes particles and some contaminant films [Asano et al. (1976)]. However, it appears to add certain undesirable contamination to the surface during the cleaning process.

UV-ozone and Other Dry-cleaning Techniques

UV radiation on surface in the presence of oxygen can remove many surface contaminants. Oxygen absorbs 185 nm radiation forming active ozone and atomic oxygen [Vig (1987)]. This technique is effective for oxidative removal of absorbed organics, but it is generally not effective for most inorganics or metals. Dry ice snow method to remove particles from wafer surface was presented in [Hoenig (1988)].

9.6 RCA Standard Clean

9.6.1 Introduction

The most standard and long-standing surface cleaning protocol of silicon wafers in the wafer manufacturing and microelectronics processing is the "RCA standard clean" process. First introduced to device fabrication in 1965 and published in the original paper by Kern in 1970 [Kern and Puotinen (1970)], the RCA standard clean process is the most cited work [Kern (1983a,b)] and practiced protocol in silicon wafer processing that stood the test of time. The RCA standard clean protocol is widely employed in wafer cleaning

"...*because extensive analytical and device reliability studies by independent researchers have confirmed the process to be the most effective cleaning method known for silicon. The process is safe and simple, has economic and ecological advantages, uses high-purity solid-free reagents, and has been used successfully in industry.*" [Kern (1983a)]

The effectiveness of the RCA cleaning method was demonstrated originally by radioactive-tracer techniques [Kern (1963a,b)], and was later confirmed by extensive analytical studies and device reliability tests by Kern and many other researchers [Kern (1983a)]. In the following sections, the RCA standard clean protocol will be described with details, along with the techniques and variations of application.

9.6.2 RCA Cleaning Protocol

The RCA Standard Clean process is based on [Kern (1983a,b)]:

1. Oxidation and dissolution of residual organic impurities and certain metal contaminants in a mixture of H_2O–NH_3OH–H_2O_2 at 75–80°C, followed by
2. Dissolution and complexing of remaining trace metals and chemisorbed ions in H_2O–HCl–H_2O_2 at 75–80°C.

This two-step RCA protocol, which has been used since the mid 1960s, known as the "SC-1" and "SC-2" to denote "standard clean 1" (step 1) and "standard clean 2" (step 2), respectively, is described in the following.

1. **SC-1, RCA standard clean 1:** Remove organic contaminants (such as grease films and photoresist residues masking the surface), expose the wafer surface and render it hydrophilic, thereby making it accessible to aqueous chemical reagents. This step achieves organic clean and particle clean. A 2–10 min clean is recommended. This process produces a thin layer of silicon dioxide of about 10 Å on the silicon wafer surface.
   The solution composition for the SC-1 mixture is listed in Table 9.3. The water used in the solution should be distilled or DI water.
2. **SC-2, RCA standard clean 2:** Remove inorganic contaminants (such as trace metals and chemisorbed ions). This step achieves ionic clean and removes the remaining trace of metallic (ionic) contaminants. A 2–10 min clean is recommended. This step also creates a thin passivating layer on the silicon wafer surface, protecting it from subsequent contamination.
   The solution composition for the SC-2 mixture is listed in Table 9.4. The water used in the solution should be distilled or deionized (DI) water.

It should be noted that the solution compositions for the SC-1 and SC-2 mixtures in Tables 9.3 and 9.4 are not critical for effectiveness of the process, as long as one operates within the prescribed range of volume ratios of 4:1:1 to 6:1:1[2]. The solution temperature can be maintained between 75 and 80°C, but should not exceed 80°C [Kern

Table 9.3 RCA Standard Clean: mixture of the solution SC-1

Composition		Volume ratio
H_2O	DI water	4~6
NH_4OH, 29%(w/w)*	Ammonium hydroxide	1
H_2O_2, 30%(w/w)	Hydrogen peroxide	1

*For example, the 29%(w/w) denotes the weight-to-weight ratio of the NH_4OH and water when preparing the solution.

---

2 A volume ratio of 5 H_2O is recommended for SC-1; while a volume ratio of 6 H_2O is recommended for SC-2 [Kern (1990)].

Table 9.4 RCA Standard Clean: mixture of the solution SC-2

Composition		Volume ratio
H_2O	DI water	4–6
HCl, 37%(w/w)	Hydrochloric acid	1
H_2O_2, 30%(w/w)	Hydrogen peroxide	1

(1983a)]. A common mistake of using the RCA clean protocol is to operate at a solution temperature exceeding 80°C, causing the reaction of the hydrogen peroxide to dissipate quickly and prematurely.

A few remarks are in order when applying the RCA clean method:

- The wafer should never be allowed to dry throughout the process of cleaning. The cleaned wafers are dried only after cooling the solution by dilution with cold water.
- Vapors of NH_3 and HCl will form a smog of NH_4Cl; thus, two separate exhaust hoods are required for SC-1 and SC-2 to avoid such chemical reaction.
- Laboratory glassware made of boronsilicate, such as Pyrex glassware, should not be used with SC-1 and SC-2 solutions because its impurity can escape and cause contamination. Instead, beakers or fused silica should be used as containers of SC-1 and SC-2 solutions.
- Excessive heating over 80°C should be avoided.
- Unstabilized H_2O_2 should be used, not the stabilized H_2O_2 with additives that are contaminants[3].

9.6.3 Techniques and Variations of the RCA Method

Initially, simple immersion technique was used for the RCA cleaning process. Alternatively, two methods were introduced [Kern (1983b)] as follows.

1. Centrifugal spray cleaning: This method was developed by the FSI Corp. to process wafers enclosed in a chamber purged with nitrogen. The process is automated to reduce the amount of chemicals used in cleaning.
2. Megasonic non-contact cleaning: Megasonic cleaning utilizes transducers which create an acoustic field at high frequency (typically 0.8–2 MHz) in a sonic bath, producing controlled cavitation on the wafer surface that achieves the effect of non-contact surface cleaning without causing surface damage. In the RCA patented process in 1975, sonic waves of 85 kHz were created by piezoelectric transducers to create cavitation on the wafer surface to remove particles down to 0.3 μm in size with input power of 5–10 W cm^{-2} [Mayer and Scwartzman (1975)].

Many variations to the ordering of the steps and chemical ratios are used throughout the industry. New solution recipes are being patented and used continually

---

3 The hydrogen peroxide must be low in aluminum and stabilizer additives, such as sodium phosphate, sodium stagnate, or amino derivatives, to prevent wafer recontamination.

although the main protocol is practically the same. In all steps, the chemicals must be electronic grade with high purity and accurate assay.

- Step 1: Use the SC-1 solution of DI water:ammonium hydroxide:hydrogen peroxide at 5:1:1 ratio (cf. Table 9.3) in a megasonic clean (75–80°C) to remove silica and silicon particles from the wafer, as well as certain organic and metal surface contamination. A clean time of 2–10 min is recommended. After cleaning, wafers must be rinsed in distilled water before the next step. This treatment creates a layer of silicon dioxide on the silicon surface and metallic contamination, typically iron.
- Step 2: This is a process of oxide removal. Use a 15–60 s dip in a solution of HF:DI water at 1:100 or 1:50 ratio at room temperature to remove the native oxide layer and any contamination in the oxide from the wafer surface. This step is optional. A strong rinse in distilled water is required after this cleaning step.
- Step 3: Use the SC-2 solution of DI water:hydrochloric acid:hydrogen peroxide at 6:1:1 (or 5:1:1) ratio (cf. Table 9.4) in a megasonic clean (75–80 °C) to remove certain ionic and metal surface contamination. A clean time of 2–10 min is recommended. After cleaning, wafers must go through a strong rinse in distilled water before the next step. This process removes metal contamination.
- Step 4: Wafers should be rinsed and dried in a standard spin-rinse dryer.

9.6.4 The Evolution of Silicon Wafer Cleaning Technology

The evolution of silicon wafer cleaning processes and technology is traced and reviewed from the 1950s to 1989 by Kern (1990). The following is a quote from the summary of the paper:

"What has changed is its implementation with optimized equipment: from simple immersion to centrifugal spraying, megasonic techniques, and enclosed system processing that allow simultaneous removal of both contaminant films and particles. Improvements in wafer drying by use of isopropanol vapor or by 'slow pull' out of hot deionized water are being investigated. Several alternative cleaning methods are also being tested, including choline solutions, chemical vapor etching, and UV/ozone treatments." [Kern (1990)]

9.7 Summary

In this chapter, several topics of wafer manufacturing are presented. Relevant to the wafer surface lapping process, wafer grinding processes are introduced for the surface processing of as-sliced wafers, especially for large crystalline wafers such as 300 mm silicon wafers, or materials with high hardness such as sapphire. Various types of grinding mechanisms and configurations are discussed, including single- and double-sided grinding processes.

Edge grinding is an important process to remove chipped edges along the peripheral edge of a silicon wafer to avoid stress concentration, crack propagation, and contamination during handling and wafer processing. SEMI defines the edge profile for silicon wafers of 300 mm or less, and 450 mm.

Etching are presented with acid etching, caustic etching, and preferential etching, with technical challenges. Various commonly used chemical etch solutions to delineate defects of silicon for analysis are presented.

After that, several topics in wafer surface cleaning are discussed. The nature of impurity and contamination on wafer surfaces is presented first. The RCA cleaning protocol is the presented with variations of RCA cleaning method, as well as the evolution of silicon wafer cleaning technology.

References

Asano M, Cho T and Muraoka H 1976 Abstract 453 *The Electrochemical Society Extended Abstract*, vol. **76-2**, p. 911, Las Vegas, NV.

Axus Technology 2013 Wafer edge grinding process (wafer edge profiling) application note. Technical report, Axus Technology.

Bardina J 1988 Methods for surface particle removal *The proceedings of the Symposium on Particles on Surfaces: Detection, Adhesion and Removal*, p. 329. Plenum Press, New York.

Brandt G 1987 Method for machining workpieces of brittle hard material into wafers. Technical report, GMN Georg Mueller Nuernberg AG. US Patent 4,663,890.

Burggraaf PS 1981 Wafer cleaning: Brushes and high-pressure scrubbers. *Semiconductor International* **4**(8), 71.

Burkman DC, Peterson CA, Zazzera LA and Kopp RJ 1988 Understanding and specifying sources and effects of surface contamination of silicon processing. *Microcontamination* **6**(11), 107–111.

Chandler TC 1990 MEMC etch-a chromium trioxide-free etchant for delineating dislocation and slip in silicon. *J. Electrochemical Society* **137**, 944.

Chapman Instrument 2013 Semiconductor wafer edge analysis note. Technical report, Chapman Instrument.

Cong WL, Zhang PF and Pei ZJ 2009 Experimental investigations on material removal rate and surface roughness in lapping of substrate wafers: A literature review. *Key Engineering Materials* **404**, 23–31.

Dash WC 1956 Copper precipitation on dislocations in silicon. *J. Applied. Physics* **27**, 1193.

Dobrescu T, Pascu N, Opran C and Bucuresteanu A 2012 Subsurface damage in grinding silicon ceramics *Annals of DAAAM for 2012 and Proceedings of the 23rd international DAAAM symposium.*

Dobrescu TG, Nicolescu AF, Pascu N and Dobre D 2011 Flattening of silicon wafers In *Annals of DAAAM for 2011 and Proceedings of the 22nd International DAAAM Symposium* (ed. Katalinic B), vol. **22**. DAAAM International, Vienna, Austria, EU.

(ed. Hull R) 1999 *Properties of Crystalline Silicon (Emis Series)*. The Institution of Engineering and Technology.

Eisele KM and Klausmann E 1984 Effects of heavy metal contamination from corrosive gas and dopant handling equipment in silicon wafer processing. *Solid State Technology* **27**(10), 177.

Gao S, Kang RK, Guo DM and Huang QS 2010 Study on the subsurface damage distribution of the silicon wafer ground by diamond wheel. *Advanced Materials Research* **126-128**, 113–118.

Hoenig SA 1988 *Particles on Surfaces 1: Detection, adhesion and Removal* Plenum Press chapter on Fine Particles on semiconductor surfaces, pp. 3–16.

Kern W 1963a Radioisotopes in semiconductor science and technology. *Semiconductor Products and Solid State Technology* **6**(10), 22–26.

Kern W 1963b Use of radioisotopes in the semiconductor field at RCA. *RCA Engineer* **9**(1), 62–68.

Kern W 1970 Radiochemical studies of semiconductor contaminations–ii, adsorption of trace impurities. *RCA Review* **31**, 234–264.

Kern W 1983a Cleaning solutions based on hydrogen peroxide for use in silicon semiconductor technology. *Citation Classic, Current Contents, Engineering, Technology, and Applied Sciences* **14**(11), 18.

Kern W 1983b Hydrogen peroxide for silicon wafer cleaning. *RCA Review* pp. 21–25.

Kern W 1990 The evolution of silicon wafer cleaning technology. *Journal of Electrochemical Society* **137**(6), 1887–1892.

Kern W and Deckert CA 1978 Chemical etching In *Thin Film Processes* (ed. Vossen JL and Kern W) Academic Press New York chapter V-1, p. 411.

Kern W and Puotinen D 1970 Cleaning solutions based on hydrogen peroxide for use in silicon semiconductor technology. *RCA Review* **31**, 187–206.

Kern W and Schnable GL 1987 Wet etching In *The Chemistry of the Semiconductor Industry* (ed. Moss SJ and Ledwith A) Chapman and Hall New York chapter 11, pp. 225–280.

Khilnani A 1988 Cleaning semiconductor surfaces: Facts and foibles In *The proceedings of the Symposium on Particles on Surfaces: Detection, Adhesion and Removal* (ed. Mittal KL), pp. 17–35. Plenum Press, New York.

Koyo Machine Industries 2019 Grinding machines for semiconductor wafers by the Koyo Machine Industries and Krystec Technology https://www.crystec.com/kmisemie.htm.

Kulkarni MS and Erk HF 2000 Acid-based etching of silicon wafers: Mass-transfer and kinetic effects. *Journal of The Electrochemical Society* **147**(1), 176–188.

Li ZC, Pei ZJ and Fisher GR 2006 Simultaneous double side grinding of silicon wafers: a literature review. *International Journal of Machine Tools and Manufacture* **46**(12-13), 1449–1458.

Licciardello A, Puglisi O and Pignataro S 1986 Effect of organic contaminants on the oxidation kinetics of silicon at room temperature. *Appl. Phys. Lett.* **48**(41), 41–43.

Liu JH, Pei ZJ and Fisher GR 2007 Elid grinding of silicon wafers: a literature review. *International Journal of Machine Tools and Manufacture* **47**(3-4), 529–536.

Liu W, Pei ZJ and Xin XJ 2002 Finite element analysis for grinding and lapping of wire-sawn silicon wafers. *Journal of Materials Processing Technology* **129**(1-3), 2–9.

Lu WK, Pei ZJ and Fisher GR 2007 A grinding-based manufacturing method for silicon wafers: decomposition analysis of wafer surfaces. *Machining Science and Technology* **11**(1), 81–97.

Mayer A and Scwartzman S 1975 Megasonic cleaning system.

Menon VB, Michaels LD, P. Donovan R and Ensor DS 1989 Effect of particulate size, composition, and medium on silicon wafer cleaning. *Solid State Technology* **32**(3), S7–S12.

Moldovan N and Ilie M 1996 From atomic parameters to anisotropic etching diagrams. *Mater. Sci. Eng.* **B37**, 146–149.

Monkowski JR 1987 Particulate surface contamination and device failures In *Treatise on Clean Surface Technology* (ed. Mittal KL) vol. 1 Plenum Press New York chapter 6, pp. 123–148.

Nakano M, Uchiyama I, Ajito T and Kudo H 2000 Method of purifying alkaline solution and method of etching semiconductor wafers. Technical report, Shin-Etsu Handotai Co., Ltd. US Patent 6,110,839.

Nishiguchi M, Sekiguchi T, Miyoshi I and Nishio K 1991 Surface grinding machine. Technical report, Asahi Diamond Industrial Co. Ltd., Sumitomo Electric Industries Ltd., and Nissei Industry Corp. US Patent 5,035,087.

Pei ZJ and Fisher GR 2001 Surface grinding in silicon wafer manufacturing *2001 NAMRC XXIX, MR01-271* SME.

Pei ZJ, Fisher GR and Liu J 2008 Grinding of silicon wafers: a review from historical perspectives. *International Journal of Machine Tools and Manufacture* **48**(12-13), 1297–1307.

Ruzyllo J 2010 Semiconductor cleaning technology: Forty years in the making. *The Electrochemical Society Interface.*

Schimmel DG 1979 Defect etch for < 100 > silicon evaluation. *J. Electrochemical Society* **126**, 479–483.

Schimmel DG and Elkind MJ 1978 An examination of the chemical staining of silicon. *J. Electrochemical Society* **125**(1), 152–155.

Schmidt PF and Pierce CW 1981 A neutron activation analysis study of the sources of transition group metal contamination in the silicon device of manufacturing process. *Journal of Electrochemical Society* **128**(3), 630–637.

Secco d'Arragona F 1972 Dislocation etch for (100) planes in silicon. *J. Electrochemical Society* **110**, 948.

SEMI M1 2017 Specification for polished single crystal silicon wafers (M1-0918) website. URL http://www.semi.org, https://www.semiviews.org.

SEMI M73 2018 Test method for extracting relevant characteristics from measured wafer edge profiles (M73-1013) website. URL http://www.semi.org, https://www.semiviews.org.

SEMI MF1725 2010 Practice for analysis of crystallographic perfection of silicon ingots (MF1725-1110) website. URL http://www.semi.org, https://www.semiviews.org.

SEMI MF1726 2010 Practice for analysis of crystallographic perfection of silicon wafers (MF1726-1110) website. URL http://www.semi.org, https://www.semiviews.org.

SEMI MF1727 2010 Practice for detection of oxidation induced defects in polished silicon wafers (MF1727-1110) website. URL http://www.semi.org, https://www.semiviews.org.

SEMI MF1809 2010 Guide for selection and use of etching solutions to delineate structural defects in silicon (MF1809-1110) website. URL http://www.semi.org, https://www.semiviews.org.

SEMI MF1810 2010 Test method for counting preferentially etched or decorated surface defects in silicon wafers (MF1810-1110) website. URL http://www.semi.org, https://www.semiviews.org.

Semitracks, Inc. 2020 Deprocessing and silicon etch website. URL https://www.semitracks.com/reference-material/failure-and-yield-analysis/failure-analysis-die-level/deprocessing.php.

Shwartzman S, Mayer A and Kern W 1985 Megasonic particles removal from solid-state wafers. *RCA Review* **46**,81.

Sirtl E and Adler A 1961 Chromic acid-hydrofluoric acid as specific reagents for the development of etching pits in silicon. *Z. Metalkunde* **52**,529.

Skidmore K 1987 Cleaning techniques for wafer surfaces. *Semiconductor International* pp. 80–85.

Slusser GJ and MacDowell L 1987 Sources of surface contamination affecting electrical characteristics of semiconductors. *J. Vacuum Science Technology A* **5**(4), 1649–1651.

Sopori BL 1984 A new defect etch for polycrystalline silicon. *J. Electrochemical Society* **131**, 667–672.

Stacy WT, Allison DF and Wu TC 1981 The role of metallic impurities in the formation of haze defects In *Semiconductor Silicon 1981, The Electrochemical Society Softbound Proceedings Series* (ed. Huff HR, Kriegler RJ and Takeishi Y), vol. PV81-5, pp. 344–353, Pennington, NJ.

Stinson MK, Erk HF and Zhang G 2011 Silicon wafer etching compositions. Technical report, MEMC Electronic Materials, Inc. US Patent 7,938,982 B2.

Sun W, Pei Z and Fisher G 2004 Fine grinding of silicon wafers: a mathematical model for the wafer shape. *International Journal of Machine Tools and Manufacture* **44**, 707–716.

SVM 2018 Silicon valley microelectronics https://www.svmi.com/wafer-services/wafer-thinning.

Tabuchi S 1984 Grinding machine. Technical report, Disco Corp. US Patent 4,481,738.

Válek L and Šik J 2012 *Defect Engineering During Czochralski Crystal Growth and Silicon Wafer Manufacturing* Modern Aspects of Bulk Crystal and Thin Film Preparation InTech (www.intechopen.com) chapter 3, pp. 43–70.

Vandamme R, Xin YB and Pei Z 2000 Method of processing semiconductor wafers. Technical report, SunEdison Semiconductor Ltd. US Patent 6,114,245.

Vig JR 1987 *Treatise on Clean Surface Technology* Springer chapter on UV/Ozone Cleaning of Surfaces.

Wang L, Hu Z, Yu Y and Xu X 2018 Evaluation of double-sided planetary grinding using diamond wheels for sapphire substrates. *Crystals* **8**(262), 1–13.

Wolters P 2010 Innovative silicon wafer grinding technology produces 5x flatter wafers https://investor.lamresearch.com/news-releases/news-release-details/peter-wolters-innovative-silicon-wafer-grinding-technology.

Wright-Jenkins M 1977 A new preferential etch for defects in silicon crystals. *J. Electrochemical Society* **124**,757.

Yamazaki J 2006 Introduction of wafer surface grinding machine model GCG300. Technical report, Komatsu NTC Ltd.

Yang KH 1984 An etch for the delineation of defects in silicon. *J. Electrochemical Society* **131**, 1140.

Young HT, Liao H and Huang H 2006 Surface integrity of silicon wafers in ultra precision machining. *International Journal of Advanced Manufacturing Technology* **29**(3), 372–378.

10

Wafer Metrology and Optical Techniques

10.1   Introduction

The developments and applications of wafer metrology have been to meet the demand for production control for increasingly complex structures on larger wafer substrates. As electronic devices are getting smaller each day, the demand for lower cost and high performance of prime silicon wafers continue to grow. This is also the case for other semiconductor and optoelectronic wafers. The parameters that affect prime wafer manufacturing process need to be monitored carefully in order to produce wafers with extreme flatness and parallelism in a desired thickness. Measurement performance involves the following four factors for modern wafer metrology: (1) sensitivity, (2) repeatability, (3) accuracy, and (4) possibility of integration in the wafer manufacturing process. The automation of manufacturing to reduce cost also requires metrology equipment to keep up with on-line control and in situ measurement.

10.2   Evaluation and Inspection of the Wafer Surface

Silicon wafers are measured and qualified at different stages of the surface manufacturing process to identify defects and evaluate quality. This is done to eliminate unsatisfactory wafer materials from the process stream and to sort the wafers into batches of uniform thickness at a final inspection stage. These wafers will become the basic raw material, called prime or premium wafers, for new integrated circuits [Bates (2000)].

10.2.1   Wafer Surface Specifications

Surface properties of wafer are characterized by several important specifications which will be discussed in the following. As the size of the wafer increases over the years, new specifications are needed to characterize wafers and/or replace the existing specifications. Refer to Sections 1.4 and 1.5 for more detail on the discussion of

Wafer Manufacturing: Shaping of Single Crystal Silicon Wafers,
First Edition. Imin Kao and Chunhui Chung.
© 2021 John Wiley & Sons Ltd. Published 2021 by John Wiley & Sons Ltd.

surface properties of wafers and other quality measurements of wafers. As a case in point, TTV of less than 15 μm is demanded by the industry for a 18 μm IC process. Furthermore, the surfaces should be smooth (with a surface roughness of $R_a < 10$ nm) and have minimum subsurface damage less than 10 μm before the final etching and polishing. The end product should have crack-free mirror surfaces with a micro-roughness less than 1.8 Å [Young et al. (2006)].

The following specifications of wafer surface are discussed.

Thickness and TTV

The total thickness variation (TTV) specifies the difference between the minimum and maximum thickness of a wafer, and can be calculated using Equation (1.5). Note that TTV is the same as GBIR (see Section 1.4.4). Refer to Section 1.4.1 for detailed discussions on the thickness and TTV. An example of calculating TTV is illustrated in Example 1.4.1 in Section 1.4.1.

Bow and Warp

Bow and warp are two important specifications in determining the quality of a wafer at final inspection.[1] Bow is the measurement of how concave or convex the deformation of the median surface of the wafer at the centerpoint, independent of any thickness variations. Warp is the difference between the maximum and minimum deviations of the median surface relative to the backside reference plane. Equations (1.6) and (1.7) in Section 1.4.1 are used to calculate bow and warp. Refer to Section 1.4.1 for more detail on the wafer surface specifications of TTV, warp, and bow. Additional discussions on warp can be found in Section 1.4.2.

The measurements of TTV, warp, bow are performed by automated process and equipment (see also Section 1.4.3). Typical instruments used to measure wafer warp include the ADE Ultra Gage (model 9500, 9700, and 9900), KLA-Tencor, Kuroda Precision Industries Nanometro series, E+H MX 7012 gauge, and many others.

Flatness

Wafer flatness directly impacts device line-width capability, process latitude, yield, and throughput [Kulkarni and Desai (2001); Oh and Lee (2001)]. As the feature size of semiconductor devices reduces, the requirements on wafer flatness have become more stringent. In addition, the specifications for larger wafers to describe the flatness are different from those for the smaller wafers in the early days. The wafer flatness is defined in a SEMI document and characterized by four-character acronyms, with flatness determination performed in five steps (cf. Table 1.5). An example of GBIR is given in Example 1.4.2. Refer to Section 1.4.4 for more detail, with reference to various SEMI documents.

Nanotopography

As the integration level of semiconductor devices increases with more and more layers being lithographically etched or deposited onto the wafer surface, nanoto-

---

1 For example, a normal or good bow on an 200 mm prime wafer would be less than 30 μm; while a good warp value on a 200 mm prime wafer would be less than 20 μm.

pography has become very important [Bhagavat and Kao (2006); Bhagavat et al. (2010)]. Nanotopography is defined as the deviation of the wafer front surface within a spatial wavelength range of approximately 0.2–20 mm (see Figure 1.9), and determines the uniformity of chemo-mechanical planarization used in processing submicron microelectronic multi-layer devices [SEMI M43 (2018); SEMI M78 (2018)]. Nanotopography is also called nanotopology. The wafer nanotopography specifications from the semiconductor industry have become more stringent over the years, especially with the increase of wafer size. Refer to Section 1.4.5 for more information.

Note that nanotopography differs from flatness defined previously. As an example, if the front and back surfaces of a wafer are piecewise parallel to each other, but the wafer has identical surface irregularities on the front and back surfaces, this wafer will be considered perfectly flat (SFQR = 0) but this wafer will exhibit nanotopography.

Surface Roughness Measurement

Surface roughness consists of fine irregularities resulting from production processes [Drozda and Wick (1983)]. Surface roughness parameters include amplitude parameters of heights, spacing parameters, and hybrid parameters [Gadelmawla and Koura (2002)]. The topic of surface roughness was presented and discussed in Section 1.4.6 with conventional definitions of several surface roughness measures, as well as the specifications of surface roughness on polished wafers published by SEMI. A seven-step procedure for roughness measurement specification of planar surfaces on polished wafers is stipulated by SEMI [SEMI M40 (2014)] for describing the measurement of roughness with elements and codes. See also Example 1.4.3 of a roughness measure of wafer surface based on the specifications of measurement using the seven-step procedure with the reporting notation and related output.

Surface roughness can be measured with a variety of instruments. These instruments include Daktak stylus profilers, Wyko optical profilers from Veeco, Surfanalyzer 5600 from Mahr Federal Inc., P-6 stylus profiler from KLA-Tencor, Z3D-720 metrology system from Zygo and other Zygo surface measurement equipment, atomic force microscope (AFM) suppliers, and many others. A typical output of a Zygo surface measurement equipment is shown in Figure 10.1 to measure an as-sliced 150 mm silicon wafer sliced by a slurry wiresaw. The equipment has both measurement and analysis, showing microscopic application and providing surface map, intensity map, and surface profiles.

Others

Other considerations of wafer surface include crystalline orientation, resistivity, wafer thickness variation (ΔTHK), rotational asymmetry (ΔROT), and others, although these are not as often used to describe wafer surface as the ones presented earlier.

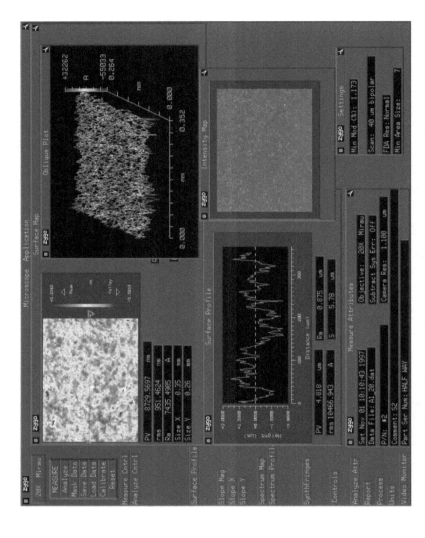

Figure 10.1 Surface measurement of a 150 mm silicon wafer sliced by a slurry wiresaw using a Zygo surface measurement equipment. The output shows microscopic application and provides measurement results in surface map, intensity map, and surface profiles. Source: Imin Kao.

10.3   Wafer Defects and Inspection

Prime wafers that meet all the specifications are packaged for device fabrication. Silicon wafers can have different defects resulting from the many processes of wafer manufacturing, especially during crystal growth in which interruptions of regular single crystalline patterns occur with imperfection. Many equipment and processes were established to delineate defects, for example, through the various methods of preferential etching in Section 9.4.3.

This section deals with three main themes about wafer defects and inspection in wafer manufacturing, including defect classification, impacts on device yield and performance, and defect inspection techniques an systems.

10.3.1   Defect Classification

Crystal defects are usually classified according to their shape and dimension [Válek and Šik (2012)]. Most defects are induced by manufacturing processes, including a plethora of processes in wafer manufacturing, from crystal growth to the cleaning of polished wafers (cf. Chapter 2 and Figure 2.1). The defects discussed here are classified as: point defects, line defects, planar defects, bulk defects, and defects on the wafer.

Point Defects

Point defects are the ones that occur only at or around a single lattice point. A crystal is said to have point defects when a silicon atom is missing in the crystal lattice structure, in particular during the process of crystallization such as the crystal growth process. A missing silicon atom in the structure leads to vacancies and formation of small clusters of vacancies and interstitials. These are also called the intrinsic point defects and can occur during the process of melt silicon and crystal growth when a silicon atom is missing, leading to a double bond forming between two nearby silicons. These vacancies are also called silicon self-interstitials [Pichler (2004); Válek and Šik (2012)].

Extrinsic point defects are primarily caused by doping process in which common dopant species, such as boron, arsenic, antimony, and phosphorous, are introduced into the silicon melt during the growth process [Pichler (2004)]. Such a doping process is necessary for silicon to possess semiconductor properties. Extrinsic defects are also called substitutional defects. The substitutional atom is referred to as an impurity. However, undesirable extrinsic point defects can be caused by the diffusion of the metallic atoms in the middle of the periodic table, such as copper, nickel and iron [Graff (2000)]. These metallic elements are the contaminants to be avoid in surface processing of silicon wafers.

Other atoms exist during crystal growth may also cause extrinsic point defects. Carbon is an unavoidable impurity in crystal growth process because of the graphite element introduced in the crucible (cf. Figure 2.2). The carbon concentration in semiconductor silicon is typically below 5×10^{15} cm^{-3}; that is, 0.1 parts per million atomic (ppma). This concentration of carbon in silicon is low and does not have a major influence [Válek and Šik (2012)]. Even with the negligible influence of carbon at low concentration and being electrically inactive, the presence of substitutional carbon can cause stress that is observable by X-ray topography.

In addition, another extrinsic defect is from oxygen which is incorporated during the crystal growth process. Oxygen atoms occupy interstitial positions in the silicon lattice and can be connected through a covalent bond to the two nearest silicon atoms. Interstitial oxygen may become "thermal donors" when electrically active chains are formed. Oxygen also forms complexes with intrinsic point defects [Pichler (2004)]. These complexes may agglomerate to form bulk defects and promote oxygen diffusion [Válek and Šik (2012)].

Line Defects

Dislocations are linear defects around which the atoms of the crystal lattice are misaligned, including the edge and screw dislocations [Hirth and Lothe (1992)]. Dislocations can move if the atoms from one of the surrounding planes break their bonds and re-bond with the atoms at the terminating edge, typical under the situation of external stress.

Line defects of edge and screw dislocations in silicon usually happen under the conditions of high stress, especially under the high temperature and high stress created in manufacturing and fabrication processes.

Planar Defects

Planar defects include grain boundaries, antiphase boundaries, stacking faults, twin boundary, and steps between flat terraces of single crystals. Stacking faults are common in silicon and other closely packed structures, and include two types:

1. Intrinsic stacking faults: formed when atomic planes are missing.
2. Extrinsic stacking faults: formed when there are additional atomic planes.

Silicon always exhibits the extrinsic stacking faults [O'Mara (1990)]. Stacking faults originate from the condensation of silicon self-interstitials [Ravi and Varker (1974)], preferentially on suitable nucleation sites such as oxide precipitates [O'Mara (1990)], metal precipitates or locations with mechanical damages in the lattice. Stacking faults can be delineated by preferential etching (cf. Section 9.4.3) of the wafer surface and examined under various microscopy techniques [Válek and Šik (2012)].

Bulk Defects

Bulk defects are three-dimensional macroscopic defects that includes (i) voids, which are a cluster of vacancies, and (ii) precipitates, which are a different phase formed in a cluster with impurities.

Bulk defects in silicon have different types, including vacancy-type defects, interstitial-type defects, and oxide precipitates [Válek and Šik (2012)]. The vacancy-type defects can be classified as void of size about 10 nm. A void intersected by the wafer surface creates a pit referred to as the "crystal-originated particle" (COP). Interstitial-type defects form when silicon self-interstitials coalesce into A and B defects (swirls). Oxide precipitates (or oxygen precipitates) can be formed in the crystal growth process due to the precipitation of oxygen interstitials. Oxide precipitates are amorphous SiO or SiO_2. Growth of oxide precipitates is usually accompanied by the formation of extended defects such as stacking faults and punched-out dislocation loops. More detail of bulk defects and related subjects can be found in Borghesi et al. (1995); Föll, H. and Kolbesen, B. O. (1975); Itsumi (2002); J. Vanhellemont and S. Senkader and G. Kissinger and V. Higgs and M.-A. Trauwaert and D. Graef and U. Lambert and P. Wagner (1997); Ryuta et al. (1990); Shimura (1994); Válek and Šik (2012); Yamagishi et al. (1992).

Defects on a Wafer

Defects on a wafer can be caused by point defects, including silicon self-interstitials, vacancies, interstitial impurities, and extrinsic point defects from substitutional impurities such as dopants and carbon, as discussed earlier. Such defects can also take the form of line defects, including edge dislocations, screw dislocations, and dislocation loops, with which a dark area will be shown on the wafer. Planar defects, including stacking faults, can nucleate through scratch on the surface of silicon wafers. The bulk defects can be a result of agglomerates of point defects.

In addition to these defects, haze with very small metal precipitation, as well as tweezer mark and/or scratches can appear on the surface of wafer as defects.

10.3.2 Impact of Wafer Defects on Device Yield and Performance

The defects in silicon wafers usually have adversely negative effects on the yield and performance of the fabricated electronic devices on the wafers. For example, a COP pit defect can cause dielectric breakdown of MOS capacitors and failure of DRAM [Yamagishi et al. (1992)]; dislocation cores are sources of shorts and leakage currents; oxide precipitates can cause an increase in leakage currents of the p–n junctions, a decrease in refresh time in DRAM memories, and a decrease in breakdown voltage in bipolar devices [Schöder (1989)].

However, not all defects are bad. The presence of oxygen and oxide precipitates can be beneficial to certain manufacturing operations by preventing the formation of process-induced defects. An example is the presence of oxide precipitates, which can act as gettering sites for contamination. Thus, the oxide precipitates in silicon wafers may getter undesirable contaminants. The contamination is usually metallic impurities, introduced during the various processes in fabrication of circuits and devices, such as the contamination contents of metal in various etchants. If not gettered, the contamination can reduce device manufacturing yields and degrade device or circuit performance. Thus, the oxygen precipitation characteristics of the silicon wafer can significantly affect both yield and performance [SEMI MF1239 (2016)]. Oxygen is introduced into silicon wafers during the crystal growing process. Hence, it is important to control the oxygen content of silicon crystal growth. The oxide precipitates are stable defects and have positive impacts through mechanical strengthening, impurity hardening, and intrinsic gettering [Rozgonyi et al. (1976); SEMI MF1188 (2018); Sumino et al. (1980); Válek and Šik (2012)].

Carbon has been suggested to play a role in defect formation, such as swirl, and has been shown to be related to the nucleation for the precipitation of oxygen. Substitutional carbon, on the other hand, can cause mechanical stress in the single silicon crystals that can be observed by X-ray topography. In addition, carbon as a contaminant has been associated with (i) the reverse bias characteristics of power devices, and (ii) the annealing problem in neutron transmutation doped silicon [SEMI MF1391 (2012)].

Most crystal defects are created during the crystal growth process, and some during the other wafer manufacturing processes. Such defects can have fundamentally damaging effects on device fabrication when the prime wafers are used in microelectronics fabrication. As the device feature size continues to shrink with more complex fabrication processes, it is ever more important to understand the defects on wafers and to study the means to reduce the defects on crystals. Such needs demand the continuous improvement of the technology of silicon crystal growth and wafer manufacturing.

10.3.3 Defect Inspection Techniques and Systems

Because defects on wafers have such a profound impact on the yield and performance of the final outcome of IC fabrication, it is very important to inspect and detect such defects for further study and investigation. Various techniques and systems are commonly used as analytical tools to detect defects in wafers and identify their location for failure analysis. Some references regarding the principles of defect inspection systems and techniques specifically relevant to silicon can be found in Keefer et al. (2002); Válek and Šik (2012).

A few techniques and systems for inspecting and detecting common defects in silicon are discussed in the following.

Preferential Etching
Preferential etching is a simple and yet effective technique to delineate defects in silicon wafers, to evaluate wafer quality, and to initiate failure analysis of an errant device. Different etchants for the preferential etching are discussed in Section 9.4.3, including the use of etchants of Dash, MEMC, Schimmel, Secco, Sirtl, Sopori, Wright, and Yang. Several defects were illustrated with microscopy photos in SEMI MF1809 (2010), such as oxidation stacking faults, shallow pits (haze), dislocations, slip dislocations, oxidation induced stacking faults, epitaxial stacking faults, damage induced slip dislocations, flow pattern defects, bulk oxidation stacking faults, scratch induced oxidation stacking faults, damage induced oxidation stacking faults, and etching stain artifacts. The photos are provided at different magnification and/or with different manifestations for some of the defects among the list of defects.

X-ray Topography
Diffraction topography, based on Bragg diffraction, can provide images that record the intensity profile of a beam of X-rays diffracted by a crystal. X-ray topography

represents a two-dimensional spatial intensity mapping that reflects the distribution of scattering power inside the crystal which reveals the irregularities in a non-ideal crystal lattice. Although the study of X-ray diffraction started in the 19th century, X-ray topography benefited from the introduction of synchrotron light sources in the 1970s, which provides considerably more intense X-ray beams to deliver better contrast and higher spatial resolution. This makes X-ray topography a powerful tool for the inspection of defects in silicon crystals and other semiconductor crystals.

When applied to the detection of silicon defects, an incident X-ray beam on a silicon surface is reflected by the atomic planes of the crystalline silicon. Constructive interference occurs when the Bragg condition is met with spatial intensity mapping of the diffraction pattern recorded on the detector. Irregularities of the non-ideal crystal lattice, such as the various defects and stains, are captured through the distribution of scattering power inside the crystal as a distortion of the image of the perfect crystal. X-ray topography can detect irregularities such as phase boundaries, defective areas, cracks, scratches, growth striations, as well as most of the common silicon crystal defects such as dislocations, oxide precipitates, stacking faults, and interstitial-type defects [Válek and Šik (2012)].

Infrared Absorption Spectroscopy

Spectroscopic techniques can be employed to identify and study chemical substances. The technique of infrared spectroscopy is conducted with an instrument called an infrared spectrophotometer (or spectrometer) to produce an infrared spectrum. Molecules will absorb frequencies of radiation when the frequency is the same as the characteristic of vibration frequency of their structure – called the resonant frequency. Infrared absorption spectroscopy exploits the fact of resonant absorptions and visualizes the results in a graph of infrared light absorbance on the vertical axis versus frequency or wavelength on the horizontal axis. The term of frequency used in infrared spectra is "wave number" which is the reciprocal of centimeter and has a unit of cm^{-1}.

The infrared spectrophotometer used for the detection of oxygen contents in silicon can be a computer assisted dispersive infrared (DIR) absorption spectrophotometer or a Fourier transform infrared (FT-IR) absorption spectrophotometer. The required resolution of the infrared absorption spectrophotometer is 4 cm^{-1} or better for the FT-IR absorption spectrophotometers, and 5 cm^{-1} or better for the DIR absorption spectrophotometers.

The FT-IR absorption spectrophotometer is most commonly used for determining the concentration of oxygen in silicon materials. Undoped silicon is transparent to IR radiation. Impurities will cause the absorption of radiation in specific frequencies owing to the lattice vibrations. The absorption due to the anti-symmetric vibrational mode of oxygen-in-silicon band is at the wave number of 1107 cm^{-1} which is used for determining the interstitial oxygen content of the silicon slice, with a suitable set of reference materials after calibration for comparison. Because the wave number of the oxygen-in-silicon band is 1107 cm^{-1}, the infrared transmittance spectrum is configured to measure over the range of wave numbers from 900 to 1300 cm^{-1} [SEMI MF1188 (2018)]. The measurement of carbon in silicon can also be accomplished at a temperature lower than 15 K using the low temperature Fourier transform infrared (LTFT-IR) spectrophotometer. At this low temperature, the detector has a higher signal-to-noise ratio in addition to a narrower full width half maximum (FWHM) absorption band [SEMI MF1630 (2018)].

Other Techniques

In addition to the three most common techniques and systems for the detection of silicon defects in the preceding discussions, namely, the preferential etching, the X-ray topography, and the infrared absorption spectroscopy, there are other techniques for silicon defect detection. They are briefly discussed in the following. The readers may want to refer to Válek and Šik (2012) for more detail.

Precipitation test: The precipitation test is used to assess the content of oxygen precipitation in wafers, after being subjected to a standardized thermal treatment sequence. The extent of oxygen precipitation is evaluated by comparing the oxygen concentrations both before, and after, the thermal treatment. The contents of the oxygen precipitation can be measured by FT-IR absorption spectrophotometer, described earlier, or by the "cleave-and-etch" analysis[2].

Oxidation induced stacking fault (OISF) test: The OISF test is based on the behavior that silicon interstitials injected below the wafer surface during oxidation under usual conditions needs a suitable nucleation site for the formation of OISFs [Ravi and Varker (1974)]. The nucleation sites can arise from the subsurface damage caused by the wafer manufacturing processes (such as sawing, grinding and lapping processes; cf. Section 2.5), contamination, or defects such as dislocations and oxygen precipitates. Thus, this method can also be used to evaluate the quality of wafer surface in general. The test wafer is first subjected to an oxidation cycle, then the oxide layer is stripped with HF acid, and finally the surface is preferentially etched. The stacking faults observed on the surface decorate the nucleation centers in the near-surface wafer region. Examples of (100) and (111) silicon wafers subjected to 8 h oxidation at 1100 °C, followed by the modified Dash preferential etch for the removal of a 4–5 μm surface layer, are illustrated with photos in SEMI MF1809 (2010).

Crystal-originated particle (COP) test: The RCA SC-1 etch (cf. Section 9.6) can delineate the COPs with the particle inspection methods [Ryuta et al. (1990)]. When etched by SC-1 chemicals, the COPs are enlarged, enabling the instruments to detect defects. The size of COP defects after the SC-1 treatment is typically 100–300 nm and the surface density is of the order of 10–100 cm^{-2} [Válek and Šik (2012)].

10.4 Measurement of the Wafer Surface Using Moiré Optical Metrology

Measuring the wafer surface to obtain the TTV and waviness (cf. Sections 1.4.1) is very important to quantify the quality of as-sliced wafers after the wiresaw

---

2 Under the cleave-and-etch analysis, the wafer cross-section is prepared by cleaving or grinding, and the sample is preferentially etched. After that, the defect distribution and density can be evaluated under the microscope.

machining process, as well as polished wafers. A standard in the industry is to use measuring equipment that utilizes a pair of capacitive probes to sample discrete points on a rotating wafer [Poduje and Baylies (1988)]. This type of equipment is widely used in the semiconductor industry against which other equipment is compared. However, such equipment using capacitive probes requires the wafers being rotated with added dynamic stress.

In the following sections, an alternative optical technique is presented to perform full-field and whole-wafer measurements – the shadow moiré technique. The shadow moiré method has distinct advantage over the existing capacitive measuring device. It does not require the spinning of wafers, hence reduces undesirable inertia load which may distort the wafer surface during the process of measurement. Moreover, it can measure wafers of any size as long as the reference grating is large enough to cover the entire surface area. It can also be employed on any geometry of wafers, not simply the wafers with a round and symmetric geometry[3]. Shadow moiré methodology can be supplemented by the "phase shifting" technique to enhance the resolution of measurement by two orders of magnitude. When phase shifting is employed, the shadow moiré method can be used to measure lapped or polished wafer surface, in addition to as-sliced wafers after wiresawing processes, because of its high resolution. In addition, the shadow moiré method can be used to determine the internal stress by measuring the fringe patterns before and after annealing the wafer (annealing relieves residual stress).

10.4.1 Measurement of the Wafer Surface Using Shadow Moiré with the Talbot Effect

The word moiré means silk in French. Silk with its textile fiber arrangement can easily generate observable fringes. The moiré patterns are typically referring to the moiré fringes formed due to shape interference. For example, a TV reporter wearing a cloth with closely parallel stripes on TV can cause moiré fringe patterns to be observed by TV viewers. This is because such pattern being photographed can interfere with the shape of the light sensors to generate moiré fringes on television and digital photography. For the moiré interference fringes to appear, the two shapes (e.g., line gratings) are typically misaligned linearly or rotationally, or have slightly different pitch. Further reading on mioré interferometry can be found in references such as Chiang (1978); Patorski (1992).

Figure 10.2 illustrates moiré fringes of two identical gratings of pitch $p = 1.2$ mm with angular misalignments of 2°, 5°, and 10°, respectively. The equation of moiré fringes due to rotation is [Chiang (1978); Patorski (1992)]

$$\delta = \frac{p/2}{\sin(\alpha/2)} \approx \frac{p}{\alpha} \qquad (10.1)$$

where δ is the distance between two adjacent pale fringes (or two adjacent dark fringes), p is the pitch of the grating, and α is the angle of rotation between the two gratings in radians. Small angle approximation is assumed to render a simplified

---

3 Many photovoltaic wafers are square or rectangular, for example. Most wafers in semiconductor applications are round and axisymmetric, however.

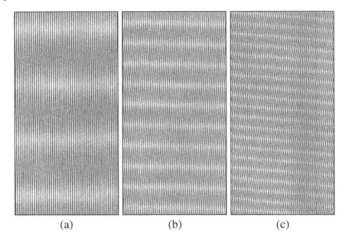

(a)   (b)   (c)

Figure 10.2 The moiré fringes resulting from rotation between two identical gratings of pitch $p = 1.2$ mm with angle of rotation (a) $2°$ $\left(\frac{\pi}{90} \text{ rad}\right)$, (b) $5°$ $\left(\frac{\pi}{36}\right)$, and (c) $10°$ $\left(\frac{\pi}{18}\right)$.

expression. Employing Equation (10.1), we find the distances between two adjacent pale fringes

$$\delta = \frac{1.2}{2 \times \pi/180} = 34.4 \text{ mm}; \delta = \frac{1.2}{5 \times \pi/180} = 13.8 \text{ mm}; \delta = \frac{1.2}{10 \times \pi/180} = 6.9 \text{ mm}$$

for $\alpha = 2°$, $5°$, and $10°$, respectively, as shown in Figure 10.2(a)–(c). The measurement of the actual size[4] is $D \approx 35, 14,$ and 7 mm, respectively.

The main interest in this section is to apply the moiré interferometry to measure the topology of wafer surfaces. The specific technique is called the shadow moiré where a glass with line gratings (called the "reference grating") is used to cast a shadow on the wafer surface to generate interference fringe patterns with the incident light source [Chiang (1969, 1978); Chiang and Jaisingh (1974); Chiang et al. (1998); Kao and Chiang (2002); Kao et al. (1998a,b); Meadows et al. (1970); Theocaris (1964a,b); Tollenaar (1945); Wei et al. (1997)]. The Talbot effect is introduced to enhance the contrast of the fringes captured by the CCD camera. The topology of the entire wafer surface can be calculated based on fringe patterns, when calibrated by the calibration specimen. The process of measurement and calculation can be automated by computers and algorithms.

An optical arrangement of the shadow moié methodology is illustrated in Figure 10.3. A point light source (beam of laser light in the figure) is directed through a reference grating, which is placed in front of the wafer surface, onto the surface of the specimen (wafer). The shadow of the reference grating on the surface of the wafer serves as the specimen grating. The reference grating and specimen grating (the shadow) will produce moiré fringes when the shadow is distorted by the variation of the depth of the wafer surface, w. Such moiré fringes containing the information of the surface topology can be captured by the digital recording

---

4 The size printed in Figure 10.2 is not the actual size due to scaling.

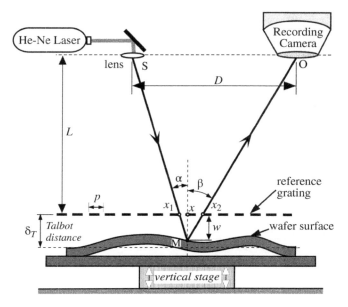

Figure 10.3 Optical arrangement of the shadow moiré method: a laser incident light is directed onto a reference grating in front of the wafer surface to produce a shadow on the wafer surface (surface profile exaggerated for illustration); the incident light and the shadow interfere with each other to generate the interferometry pattern, called "shadow moiré" which is imaged by the CCD camera, at the same level as the light source.

camera, as shown in Figure 10.3. The point light source and the camera lens are arranged at the same level parallel to the reference grating plane.

Due to the variation of depth, w, from the reference grating to the actual wafer surface, the shadow of the reference grating is distorted and will interfere with the reference grating, resulting in moiré fringes. The depth representing the surface topology of the wafer can be calculated from the following equation, assuming $D \gg Np$ [Chiang (1978), Wei et al. (1998)]

$$w = \frac{NpL}{D} = \frac{N\,p}{\tan \alpha + \tan \beta} \tag{10.2}$$

where w is the depth shown in Figure 10.3, N is the fringe order, p is the pitch of the reference grating, D is the distance between the light source and the camera, and L is the distance between the light source and the grating surface. Once the surface topology on both sides are resolved, the thicknesses of all locations on the wafer can be obtained by identifying on the surface a point of known thickness. Once the thicknesses of all points on the wafer are found, the warp and total thickness variation (TTV) can be easily calculated.

A photo of moiré fringe patterns of a single crystalline silicon wafer surface is shown in Figure 10.4, in which several fringe order can be seen from a wafer which has reasonably large surface variation. The dimensions used in the experimental setup in Figure 10.3 are:

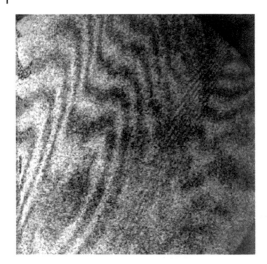

Figure 10.4 A photo with moiré fringes for the surface of a single crystalline silicon wafer using a reference grating of 1000 lines per inch, with the shadow moiré setup in Figure 10.3. See [Wei et al. (1998)] for a photo with full wafer surface. Source: Imin Kao.

- $D = 2,090$ mm
- $L = 1,400$ mm
- $p = 0.001$ in $= 0.0254$ mm
- $\lambda = 630$ nm, the wavelength of the He–Ne laser light source.

From Equation (10.2), we can calculate the depth variation between fringe orders

$$w = \frac{NpL}{D} = \frac{N(0.0254)(1400)}{2090} = (17 \times 10^{-3}) \text{ N mm} = 17 \text{ N } \mu\text{m}.$$

That is, the depth variation between each fringe order is 17 μm, using a very fine grating of 1000 lines per inch. The readers may wish to refer to Wei et al. (1998) for a more detailed treatment of the methodology.

Experimental Setup and the Talbot Effect

In order to render the required resolution of the depth, w, the reference grating must have a high spatial frequency, such as the one used in the experiment above, with 1000 lines per inch, or about 40 lines per millimeter ($p = 0.0254$ mm). However, the fringe patterns may be blurred, as a result of diffraction due to the fine pitch of the reference grating. To obtain moiré fringes with good contrast, the distance between the reference grating and the specimen must be adjusted such that the mean surface of the wafer is at a Talbot distance, which is the separation between the diffraction grating and self-image plane [Edgar (1969); Guigay (1971); Tu (1988)]. The adjustment of the Talbot distance is accomplished by a high-resolution vertical translation stage, shown in Figure 10.3, to move the wafer specimen up and down. The equation of the Talbot distance, δ_T, is given by the following equation

$$\delta_T = n \left(\frac{2 p^2}{\lambda} \right) \tag{10.3}$$

where $n = 1, 2, \ldots$ is a positive integer, p is the pitch of grating, and λ is the wavelength of the light source. The Talbot distance, δ_T, is labeled in Figure 10.3 for reference.

The light source was a He–Ne laser with a wavelength of $\lambda = 630$ nm. Assuming the same parameters of experiment setup with $D = 209$ cm, $L = 140$ cm, the shortest Talbot distance at $n = 1$ is

$$\delta_{\mathrm{T}} = 1 \left(\frac{2 \, (0.0254)^2}{630 \times 10^{-6}} \right) = 2.0 \text{ mm.}$$

During the experiments, the Talbot distance was first calculated using Equation (10.3) and then adjusted using a high-resolution vertical stage with fine adjustment, as illustrated in Figure 10.3. In this case, a translational stage was used to adjust for the Talbot distance of about 2 mm to reduce the blur and obtain a photo with sharp image of fringes for subsequent analysis.

Table 10.1 compares the capacitive and shadow moiré methods for wafer surface measurements. Some observations on the advantages of the shadow moiré technique for measuring wafer surface are offered in the following.

- The condition of wafer can be effectively detected by examining the spatial frequency of the fringes, without detailed 3D surface analysis via this whole-wafer, full-field fringe information. As a result, rejection of wafers can quickly be made based upon abnormal distribution of fringes, without very extensive computation cost.
- The wafers do not need to be spun or moved during the measuring process; therefore, it eliminates any possible distortion during the measurement, such as addition of stress due to dynamics, stress redistribution or vibration. Consequently, high fidelity of surface measurements can be obtained.
- The shadow moiré methodology can be used regardless of the shape (rectangular, round, or torus shape) and size of wafers. It can also be automated using algorithms and a computer.
- Although the distance between two neighboring fringes is 17 μ m under the experimental setup above with the prescribed parameters, the resolution of the 3D topology plots is in the order of a couple of microns with an interpolation algorithm used to calculate the fringe orders.

Table 10.1 Comparison of capacitive versus shadow moiré methods

Capacitive method	Shadow moiré method
Use capacitive probes	Use optical grating and light source
Sampling points on surface	Full-field measurement
Slow	Fast
Vacuum chuck gripper	No contact
Introduce contact stress	No stress due to contact
Dynamic, with vibration	Static, no inertial effect
Size limitation	No size limitation
Symmetric wafer geometry (round)	Any wafer geometry

- To increase the resolution, one may increase the projection and/or receiving angle, α and β as formulated in Equation (10.2). However, this might cause the size of the equipment to become too big to be practical. Alternatively, one can apply the phase shifting technique to enhance resolution. In general, a two orders of magnitude (about 100 times) increase in the resolution can be achieved. This will be discussed further in Section 10.4.2.

10.4.2   Enhancing the Resolution of Shadow Moiré with "Phase Shifting"

When using the traditional shadow moiré method, the moiré fringe patterns are analyzed by tracing the center lines of the light and dark fringes, as those shown in Figure 10.4. Data are collected only along the fringe centers. The fringe centers, however, are difficult to trace by hand or by computer, and errors are often introduced. The information of surface points between fringe centers can be obtained through interpolation. The resolution of the measurement is therefore limited by the number of fringe centers. While interpolation can create more data and generate a smooth surface, it can not enhance the physical resolution. At the same time, a great deal of supplementary information, embedded in the variation of the gray level between these light and dark fringe centers, is not used. Furthermore, additional information is needed to determine the sign of the fringe order. It is difficult to determine the relative order of successive fringes by looking at a moiré fringe pattern directly, making it difficult to distinguish which part of the wafer surface is concave or convex. To determine this, the object is moved in a known direction and the movement of fringe patterns is observed to establish the surface convexity. Typical resolution of wafer surface measurements using the conventional shadow moiré technique is in the order of 10 μm with gratings of about 1000 lines per inch and a He–Ne laser light source, as calculated in Section 10.4.1. With linear interpolation, the resolution can be improved to the order of a couple of microns. Readers may refer to Wei et al. (1998) for more detail.

In this section, a phase-shifting technique is introduced that takes advantage of the gray level information to increase the resolution of measurement. A series of fringe patterns are recorded while the fringe patterns are shifted by moving the specimen. The phase is encoded in the variations of the intensity pattern from the recorded fringe images, and a simple point-by-point calculation recovers the phase. The need to locate the fringe centers is eliminated. Similar to the discussion in Section 10.4.1, the Talbot distance is adjusted between the grating and wafer surface, using a vertical stage illustrated in Figure 10.3, such that a fringe pattern of good contrast and quality can be obtained.

Various research has been conducted using the phase-shifting technique to enhance the resolution of measurements [Creath (1985); Dirckx et al. (1988); Hu et al. (1999); Huang et al. (2002); Jin et al. (2000); Mauvoisin et al. (1994); Wei and Kao (1999); Yoshizawa and Tomisawa (1993)]. Specifically, phase shifting of shadow moiré technique was presented in Jin et al. (2000) by moving the reference grating in two directions without moving the light source, CCD camera or object. The approach presented here, however, moves the object (i.e., the wafer whose surface is to be measured) up and down within the Talbot distance using a vertical stage. The methodology of phase shifting is similar to the three-step phase-shifting algorithm in Hu et al. (1999); Huang et al. (2002) with three captured images having a phase shift of 120°, except that a four-step phase-shifting algorithm is employed here.

Figure 10.3 shows a schematic of the phase-shifting shadow moiré experiment. The light source at S passes through the reference grating at x_1, reaches the point M on the wafer surface. A part of the reflected light passes again through the reference grating at x_2, and is observed at position O. The reference grating is assumed to be sinusoidal with the following transmission function at x

$$T(x) = \frac{1}{2} + \frac{1}{2} \sin\left(\frac{2\pi}{p}x\right) \tag{10.4}$$

where p is the pitch of the reference grating. Because the light passes through the grating twice, the intensity pattern observed at O is described by

$$I(O) = H(M)T(x_1)T(x_2) \tag{10.5}$$

where $H(M)$ depends on the surface quality around the point M as well as on its position and the intensity of the light source.

Substituting Equation (10.4) into Equation (10.5) and rearranging the items, we can obtain

$$I(O) = I_1(O) + I_2(O) \tag{10.6}$$

where

$$I_1(O) = H(M)\left[1 + \frac{1}{2}\cos\frac{2\pi}{p}(x_2 - x_1)\right] \tag{10.7}$$

$$I_2(O) = H(M)\left[\sin\frac{2\pi}{p}x_1 + \sin\frac{2\pi}{p}x_2 - \frac{1}{2}\cos\frac{2\pi}{p}(x_2 + x_1)\right]. \tag{10.8}$$

From Figure 10.3, it is obvious that

$$x_2 - x_1 = \frac{Dw}{L + w}. \tag{10.9}$$

Equation (10.9) is valid under the assumption that the point light source and the camera are located at the same level parallel to the reference grating plane. When $L \gg w$, Equation (10.9) can be simplified as

$$x_2 - x_1 = \frac{Dw}{L}. \tag{10.10}$$

The phase is obtained through geometry

$$\phi = \frac{2\pi}{p}\frac{Dw}{L}. \tag{10.11}$$

Thus, $I_1(O)$ can be expressed as

$$I_1(O) = H(M)\left(1 + \frac{1}{2}\cos\phi\right). \tag{10.12}$$

The phase ϕ in Equation (10.11) is proportional to w because p, D, and L are constants. The phase angles ϕ are the same for same w. Hence, Equation (10.12), which is a function of ϕ, can be used to represent the surface contour. The intensity

$I_2(O)$ has a spatial frequency of p^{-1}. For dense grating, such as 1000 lines per inch (or about 40 lines per millimeter) used in the experiments, p^{-1} is so high that the CCD camera cannot capture it. Therefore, $I_2(O)$ in Equation (10.8) can be neglected. Thus, the intensity at O can be expressed as

$$I(O) \cong I_1(O) \tag{10.13}$$

or

$$I = I' + I'' \cos \phi \tag{10.14}$$

where I' is the average intensity, I'' is the intensity modulation [Dirckx et al. (1988)].

Similarly, the two-dimensional image captured by the CCD camera can be described as

$$I(x, y) = I'(x, y) + I''(x, y) \cos \phi(x, y). \tag{10.15}$$

Equation (10.15) is called the fundamental equation of phase-shifting technique [Malacara (1991)].

In the traditional shadow moiré method, only the extreme values of ϕ are measured by tracing the centers of the dark and light fringes. The phase difference between any two dark or light fringes is $2\pi N$, where $N = 0, 1, 2, \ldots$ is the fringe order. From Equation (10.11), the surface depth variation between any two dark or light fringe center lines becomes

$$w = \frac{NpL}{D}.$$

This equation is the same as Equation (10.2), used in the traditional shadow moiré analysis to calculate the surface depth variation at each measurement point along the center lines. Using numerical interpolation between fringes, the three-dimensional surface topology can be reconstructed (cf. Section 10.4.1). Though linear interpolation may be employed to generate a smooth contour for the entire wafer surface, it does not enhance the physical resolution of the measurement. In contrast, the phase-shifting shadow moiré method utilizes Equation (10.15) at each measurement point and recovers the phase point-by-point. Therefore, the physical resolution can be increased significantly.

In the following, the fundamental equation of phase shifting will be applied to discuss the procedures of phase wrapping and unwrapping, along with the four-step phase wrapping algorithm.

Phase Wrapping

In order to solve the phase $\phi(x, y)$, the minimum number of phase-shifting measurements of the fringe pattern required to construct the phase is three, because there are three unknowns in Equation (10.15): $I'(x, y)$, $I''(x, y)$, and $\phi(x, y)$. Various algorithms have been developed to calculate the phase [Huang et al. (2002); Malacara (1991)]. A complete study on the influence of different parameters on the phase change of the shadow moiré method was presented in Mauvoisin et al. (1994). They showed that moving the specimen perpendicularly to change its distance to the reference grating can result in a uniform phase change for the whole image. The phase variation corresponding to the change of distance can be derived from Equation (10.11)

$$\Delta\phi = 2\pi \frac{D}{pL}\Delta w \tag{10.16}$$

under the assumption of $L \gg w$, with the light source and CCD camera being at the same level.

Four-step Phase Wrapping Algorithm

In the phase-shifting shadow moiré method, it is more convenient and accurate to control the distance between the object and the reference grating. We change an equal distance at each step of phase shifting. Suppose the phase shift corresponding to each step is $\Delta\phi = 2\alpha$, where another unknowns, α, was introduced. Therefore, four equations are needed to recover the phase. The corresponding algorithm is called the *four-step phase wrapping* algorithm. The four measured fringe patterns can be represented by the following equations

$$I_1(x,y) = I'(x,y) + I''(x,y)\cos(\phi(x,y) - 3\alpha) \tag{10.17}$$

$$I_2(x,y) = I'(x,y) + I''(x,y)\cos(\phi(x,y) - \alpha) \tag{10.18}$$

$$I_3(x,y) = I'(x,y) + I''(x,y)\cos(\phi(x,y) + \alpha) \tag{10.19}$$

$$I_4(x,y) = I'(x,y) + I''(x,y)\cos(\phi(x,y) + 3\alpha). \tag{10.20}$$

The solution can be found by expanding these four equations and applying the trigonometric identity for sine or cosine of 3α [Carré (1966); Malacara (1991)] as follows

$$\phi(x,y) = \tan^{-1}\frac{\left\{[3(I_2 - I_3) - (I_1 - I_4)][(I_1 - I_4) + (I_2 - I_3)]\right\}^{\frac{1}{2}}}{(I_2 + I_3) - (I_1 + I_4)}. \tag{10.21}$$

Since $\phi(x,y)$ is proportional to the surface depth, w, as shown in Equation (10.11), the surface topology information can be derived from $\phi(x,y)$ given in Equation (10.11).

Phase Unwrapping

Because the arctangent function is defined over the range $[-\frac{\pi}{2}, \frac{\pi}{2}]$, regardless of the actual value of the phase, only values of the phase within this range will result from Equation (10.21). In order to measure the phase of more than half a wavelength, we must remove the discontinuities that occur in the phase calculation as a result of the arctangent.

The first correction made to the calculated phase is to extend the calculation range from 0 to 2π. This can be done by referring to the signs of the sine and cosine. When applying the four-step phase wrapping algorithm presented earlier, the signs of $\sin\phi$ and $\cos\phi$ cannot be detected readily. However, we can use the terms that are proportional to $\sin\phi$ or $\cos\phi$ to determine the signs relatively. One such set of terms is [Creath (1985); Malacara (1991)]

$$\sin\phi(x,y) \propto (I_2 - I_3) \tag{10.22}$$

$$\cos\phi(x,y) \propto (I_2 + I_3) - (I_1 + I_4). \tag{10.23}$$

Table 10.2   Modulo 2π phase correction

$\sin\phi(x,y)$	$\cos\phi(x,y)$	Corrected phase $\phi(x,y)$
0	+	0
+	+	$|\phi(x,y)|$
+	0	$\frac{\pi}{2}$
+	−	$\pi - |\phi(x,y)|$
0	−	π
−	−	$\pi + |\phi(x,y)|$
−	0	$\frac{3}{2}\pi$
−	+	$2\pi - |\phi(x,y)|$

With these analyses, Table 10.2 is established to convert the arctangent results to values between 0 and 2π as a function of the values of the sine and the cosine.

The next step is to remove the 2π discontinuities that are present in the raw phase data. It converts the modulo 2π phase data to a continuous representation of the phase. The sampling should satisfy the Nyquist criteria: at least two pixels between two fringe centers, and the phase change per pixel spacing should not be more than π. Therefore, the 2π discontinuities can be corrected by adding 2π or multiples of 2π to, or subtracting 2π or multiples of 2π from, the calculated value of the second pixel if the phase difference of two adjacent pixels exceeds π. The unwrapped phase value can be mapped to the surface depth variation using Equation (10.16).

Experimental Result

According to the four-step algorithm, experiments were conducted to obtain the 3D topology of the wafer surface. First, the moiré fringes for the surface of a polished specimen are recorded by a CCD camera. The results are used to measure the slope of supporting platform and to check the experiment accuracy. Then a silicon wafer sliced by wiresaw is measured and calibrated by the calibration specimen.

The phase is then unwrapped to remove the 2π discontinuities. The unwrapped values are mapped to the depth variations using Equation (10.16). This depth, however, needs to be compensated by the angle between the reference grating plane and the specimen plane on which the wafer rests. This is accomplished by computing the fringes formed on the surface of a smooth calibration specimen, which is resting on the same specimen plane.

Measurement of the Calibration Specimen

In order to measure the slope of the supporting platform on which the object is resting, a well polished calibration specimen is used. The parameters for this experiment

(a)   (b)

(c)   (d)

Figure 10.5 Measurement of a polished calibration specimen: four shifted images, each one is the average of 16 images taken at each step of the four-step phase-shifting algorithm.

setup are: $L = 11''$, $D = 76.2''$. Figure 10.5 shows the four fringe patterns for a part of the calibrated specimen surface having a dimension of 45.34 mm × 73.67 mm. The four images are recorded by a CCD camera, with the distance between the reference grating and the specimen surface increased by $\Delta w = 10 \ \mu$m, respectively. Each fringe pattern in Figure 10.5 is the average of 16 images taken at the same distance. One can see clearly that the four fringe patterns are shifted slightly from one another. The image has a width of 800 pixels and a height of 1300 pixels. The four images provide four intensity values at each pixel. Substituting the four intensity values into Equation (10.21), the phase can be calculated. Using Table 10.2, the resulting phase, ranging from $-\frac{1}{2}\pi$ to $\frac{1}{2}\pi$, can be mapped to the desired range from 0 to 2π. The results are presented in Figure 10.6. The phase map is very similar to the fringe pattern.

Figure 10.6 The phase pattern of the calibration measurement, visualizing the phase distribution on the surface, is calculated from Figure 10.5 based on the four-step phase wrapping algorithm.

The phase is then unwrapped to remove the 2π discontinuities. The unwrapped values are mapped to the depth variations using Equation (10.16). Figure 10.7 shows the 3D surface topology without calibration. The figure shows the slope of the supporting platform, with one end being about 27 μm higher than the other end. Removing the slope, we can obtain the 3D calibration specimen surface topology shown in Figure 10.8. The highest point in Figure 10.8 has a depth value of 1.31 μm, while the lowest point has a value of −0.20 μm. The difference is 1.51 μm on the surface of the calibration specimen.

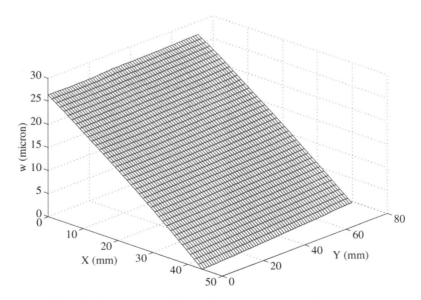

Figure 10.7 Surface topology of the calibration specimen, without compensation for the orientation of the platform on which the specimen rests.

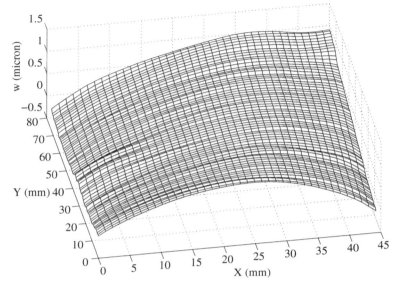

Figure 10.8 Surface topology of the calibration specimen after the orientation of the platform is accounted for. The surface variation (about 1.5 µm) is much smaller than the height due to the slope of the platform surface (about 27 µm, see Figure 10.7).

Since the difference is much smaller than the height resulting from the slope of the specimen, the slope obtained from the calibration specimen is regarded as accurate. The slope of the specimen surface in Figure 10.7 is two dimensional. With respect to the origin at $(0, 0)$, the slope along the x-direction is 0.034°, and 0.00072° along the y-direction. These calibration results are used in the measurement of wafer surface to account for the slope of the platform in order to render the true surface depth and topology of wafer surfaces.

Measurement of a Regular Wafer Sliced by a Wiresaw

Next, we conducted experiments to measure a regular silicon wafer surface, as-sliced by a wiresaw. The parameters for this experiment setup are: $L = 26''$, $D = 23''$. Similar to the procedures of measurements of specimens, four shifted fringe patterns were obtained, as shown in Figure 10.9, with the distance between the reference grating and the wafer surface increased by $\Delta w = 20$ µm. The images have a width of 800 pixels and a height of 1160 pixels, which corresponds to an area of 36 mm×52 mm. The wrapped phase pattern is visualized in Figure 10.10. After phase unwrapping and converting the phase values to the depth value, the 3D wafer surface can be reconstructed, as shown in Figure 10.11. The 3D figure shows that there is a slope from one end to the other end, as well as a sudden change of surface depth (or step) in the middle of the surface. The change of surface depth can be viewed more distinctly after flattening the surface in Figure 10.12, which shows that there is about a 10–15 µm jump resulting from a discontinuity in wiresaw slicing operation. Finally, the calibration result obtained previously can be used to

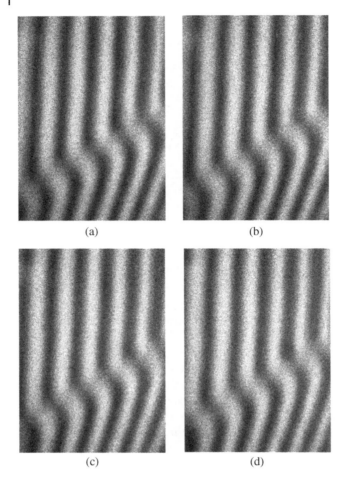

(a)   (b)

(c)   (d)

Figure 10.9 The four shifted images of a regular wafer surface from (a) to (d) correspond to Equations (10.17)–(10.20). Each of the the fringe patterns shown here is the average of 16 images taken at each step of the four-step phase-shifting algorithm.

calibrate and account for the slope of the platform. As a result, the calibrated 3D wafer surface topology is shown in Figure 10.13. Figure 10.13 shows that one end of the wafer is about 160 μm higher than the other end. This is due to the fact that one end of the wafer was deliberately made thicker than the other end for this particular experiment.

Discussions

When deriving the equations for phase-shifting shadow moiré, the second part in Equation (10.6), $I_2(O)$, is neglected due to its high spatial frequency. It has the same spatial frequency as the grating – 40 lines per millimeter. In order to capture this frequency, the sampling spatial frequency must be more than 80 per millimeter according to the Nyquist criteria. The part of the wafer under analysis has a dimension of 36 mm×52 mm with 800 × 1160 pixels. Therefore, the sampling spatial frequency

Figure 10.10 The phase pattern, visualizing the phase distribution on the wafer surface, is calculated from Figure 10.9 based on the four-step phase wrapping algorithm.

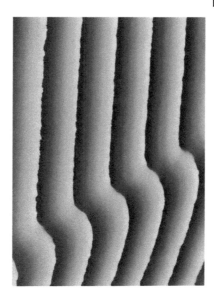

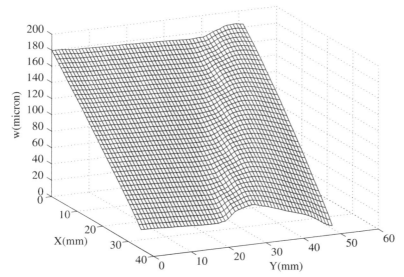

Figure 10.11 Wafer surface topology of an as-sliced wafer shown in Figure 10.9. A shift in depth can be clearly seen, as indicated by the fringe patterns in Figure 10.9. The orientation of the platform on which the wafer lies was not yet accounted for.

is 25 per millimeter, which is much less than 80 per millimeter. Thus, $I_2(O)$ can be neglected in the analysis.

The physical resolution of conventional shadow moiré method is determined by the distance represented by two consecutive fringe centers. Interpolation is required to calculate points between two fringe centers. In the phase-shifting shadow moiré method, the phase is recovered point-by-point, eliminating the need for numerical

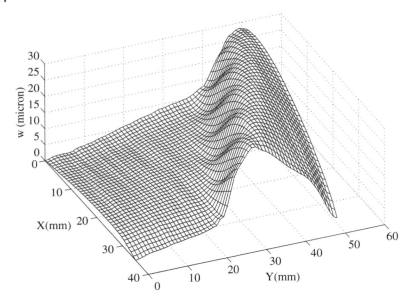

Figure 10.12 Surface topology of a wiresawed wafer after flattening the *x* and *y* directions from Figure 10.11. A sudden change of depth of about 10–15 μm can be seen.

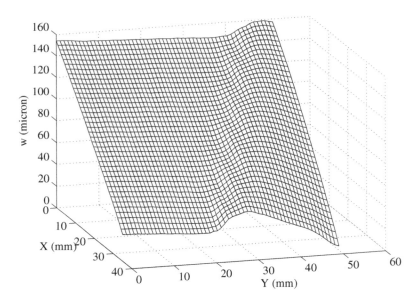

Figure 10.13 Surface topology of an as-sliced wafer after compensating for the orientation of platform.

interpolation. The physical resolution of the phase-shifting method is determined by the pixel density of the CCD camera. In this experiment, for example, there are 800 × 1160 pixels in the area of 36 mm×52 mm. There are five fringes in this region. The traditional shadow moiré method traces the five fringe centers and can provide direct information only for the points located at the centers. In contrast, the phase-shifting shadow moiré method can calculate the phases at 800 points at each row because 800 pixels are available for the calculation of the surface depth due to the intensity of the signal. Therefore, the physical resolution is increased by 160 times – a little more than two orders of magnitude.

Since the usage of fine grating introduces diffraction that blurs the fringe patterns, the identification of the Talbot effect is critical in this experiment. Although the Talbot distances are not a single distance, it has been observed that when the distance between the reference grating and the wafer surface is approximately at the first Talbot distance when $n = 1$ in Equation (10.3), the image quality is the best. The image is blurred or may even disappear altogether when the wafer is placed at distances other than the Talbot distances in Equation (10.3) away from the reference grating. In the second experiment, the four-step algorithm requires changes in the distance three times. Each time, the distance is increased by 20 μm. The total distance change is 60 μm (see Equations 10.17–10.20) with respect to the initial position. The first Talbot distance is 2 mm in our experimental setup; therefore, the images maintains good quality within the distance change of 60 μm. In general, we should ensure that the distance between the reference grating and wafer surface is always within reasonable range of the first Talbot distance.

Equation (10.3) shows that light sources of different wavelength have different Talbot distances for the same reference grating. A white light source has a spectrum, and hence does not have a single Talbot distance. The moiré fringes resulting from a white light source have color edges in experiments. The black and white CCD camera converts the color to a corresponding gray level, resulting in inaccurate data. Therefore, a He–Ne laser with a single wavelength of 630 nm was used to produce a single Talbot distance. Other monochrome light sources can also be used.

In addition, due to the speckle effect introduced by the laser, the setup should be well vibration-isolated in order to make sure the laser speckle does not move during the phase-shifting process. The static speckle will cancel each other when perform phase wrapping because of the subtraction operations in Equation (10.21), while the moving speckle will introduce artificial noise to the image.

10.4.3   Wire Web Management Using Optical Metrology Technology

Optical metrology can be applied for the inspection and management of wire web to ensure the spacing and orientation of the wire segments of the wire web are maintained within the required specification. This topic was presented and discussed in Section 5.7.2.

In addition, the online and real-time monitoring of wire wear and diameter, without taking a sample of the wire from the wiresaw machines during its operation, is important for the continuous monitoring and assessment of the integrity of the wire for slicing. An optical metrology method was presented and discussed in Section 5.7.1.

10.5   Summary

In this chapter, we first introduce the evaluation and inspection of a wafer surface, particularly the defects in the wafer and their implications to the yield and performance of the fabricated IC devices from wafers. Several defects of silicon are discussed, including point, line, planar, bulk defects, and defects on wafer surface due to processing and handling. Defect inspection techniques and systems are discussed, including the preferential etching, the X-ray topography, the infrared absorption spectroscopy (FT-IR and DIR), and others. Next, an optical metrology technique, in particular the moiré technique with the Talbot effect, is described for measuring wafer surface with a non-contact whole-wafer optical method. Phase-shifting technology is introduced to enhance the resolution of the optical moiré technique for wafer surface measurement.

References

Bates SP 2000. Technical report, Applied Materials/SJSU.

Bhagavat S and Kao I 2006 Theoretical analysis on the effects of crystal anisotropy on wiresawing process and application to wafer slicing. *International Journal of Machine Tools and Manufacture* **46**, 531–541.

Bhagavat S, Liberato J, Chung C and Kao I 2010 *Effects of mixed abrasive grits in slurries on free abrasive machining (FAM) processes.* **50**, 843–847.

Borghesi A, Pivac B, Sassella A and Stella A 1995 Oxygen Precipitation in Silicon. *Journal of Applied Physics* **77**(9), 4169–4244.

Carré P 1966 Installation et utilisation du comparateur photoelectrique et interferentiel du bureau international des poids de mesures. *Metrologia*.

Chiang FP 1969 Techniques of optical spatial filtering applied to the processing of moiré-fringe patterns. *Experimental Mechanics* **9**, 523–526.

Chiang FP 1978 Manual on Experimental Stress Analysis Society for Experimental Mechanics chapter 3.

Chiang FP and Jaisingh G 1974 A new optical system for moiré methods. *Experimental Mechanics* **14**, 459–662.

Chiang FP, Du ML and Kao I 1998 Some new applications on in-plane, shadow and reflection moiré methods *International Conference on Applied Optical Metrology*, Hungary.

Creath K 1985 Phase-shifting speckle interferometry. *Applied Optics*.

Dirckx J, Decraemer W and Dielis G 1988 Phase shift method based on object translation for full field automatic 3-d surface reconstruction from moiré topograms. *Applied Optics* **27**, 1164–1169.

Drozda TJ and Wick C 1983 Non-traditional machining vol. Vol. 1, Machining of *Tool and Manufacturing Engineers Handbook* 4th edn Society of Manufacturing Engineers (SME) Dearborn, Michigan pp. 1–23.

Edgar RF 1969 The fresnel diffraction images of periodic structures. *Optica Acta* **16**(16), 281–287.

Föll, H. and Kolbesen, B. O. 1975 Formation and Nature of Swirl Defects in Silicon. *Applied Physics A: Material Science and Processing* **8**(4), 319–331.

Gadelmawla ES and Koura MM 2002 Roughness parameters. *Journal of Materials processing Technology* **123**(1), 133–145.

Graff K 2000 *Metal Impurities in Silicon-device Fabrication* 2nd edn. Springer-Verlag Berlin, Berlin, Germany.

Guigay JP 1971 On fresnel diffraction by one-dimensional periodic objects, with application to structure determination of phase objects. *Optica Acta* **18**(9), 677–682.

Hirth JP and Lothe J 1992 *Theory of Dislocations* 2nd edn. Krieger Publishing Company.

Hu Q, Huang PS, Fu Q and Chiang FP 1999 Calibration of a 3-d surface countouring and ranging system In *Three-Dimensional Imaging, Optical Metrology, and Inspection V* (ed. Harding KG), pp. 158–166. SPIE 3835.

Huang PS, Hu QJ and Chiang FP 2002 Double three-step phase shifting algorithm. *Applied Optics* **41**(22), 4503–4509.

Itsumi M 2002 Octahedral Void Defects in Czochralski Silicon. *Journal of Crystal Growth* **237-239**(3), 263–271.

J. Vanhellemont and S. Senkader and G. Kissinger and V. Higgs and M.-A. Trauwaert and D. Graef and U. Lambert and P. Wagner 1997 Measurement, Modelling and Simulation of Defects in As-grown Czochralski Silicon. *Journal of Crystal Growth* **180**(3-4), 353–362.

Jin L, Kodera Y, Yoshizawa T and Otani Y 2000 Shadow moiré profilometry using the phase-shifting method. *Society of Photo-Optical Instrumentation Engineers* **39**(8), 2119–2123.

Kao I and Chiang FP 2002 Research on modern wiresaw for wafer slicing and on-line real-time metrology *Proceedings of NSF Design, Service, and Manufacturing Grantees and Research Conference*, San Juan, Puerto Rico.

Kao I, Prasad V, Chiang FP, Bhagavat M, Wei S, Chandra M, Costantini M, Leyvraz P, Talbott J and Gupta K 1998a Modeling and experiments on wiresaw for large silicon wafer manufacturing *the 8th Int. Symp. on Silicon Mat. Sci. and Tech.*, p. p.320, San Diego.

Kao I, Wei S and Chiang FP 1998b Vibration of wiresaw manufacturing processes and wafer surface measurement *NSF Design and Manufacturing Grantees Conference*, pp. 427–428, Monterey, Mexico.

Keefer M, Pinto R, Dennison C and Turlo J 2002 *The Role of Metrology and Inspection in Semiconductor Processing* Handbook of Thin Film Deposition Processes and Techniques: Principles, Methods, Equipment and Applications 2nd edn Noyes Publications chapter 6, pp. 241–286.

Kulkarni M and Desai A 2001 Silicon wafering process flow. US Patent 6,294,469.

Malacara D 1991 *Optical Shop Testing* 2nd edn. John Wiley & Sons.

Mauvoisin G, Brémand F and Lagarde A 1994 Three-dimensional shape reconstruction by phase-shifting shadow moiré. *Applied Optics* **33**, 2163–2169.

Meadows DM, Johnson WO and Allen JB 1970 Generation of surface contours by moiré pattern. *Applied Optics* **9**, 942–947.

Oh HS and Lee HL 2001 A comparative study between total thickness variance and site flatness of polished silicon wafer. *Japanese Jurnal of Applied Physics* **40**(1), 5300–5301.

O'Mara WC 1990 *Oxygen, Carbon and Nitrogen in Silicon* Handbook of Silicon Technology Noyes Publications Noyes Publications pp. 451–549.

Patorski K 1992 *Handbook of the Moiré Fringe Technique*. Elsevier.

Pichler P 2004 *Intrinsic Point Defects, Impurities, and Their diffusion in Silicon*. Springer-Verlag/Wien, Wien, Austria.

Poduje NS and Baylies WA 1988 Wafer geometry characterization: An overview. *Microelectronic Manufacturing and Testing*.

Ravi KV and Varker CJ 1974 Oxidation-Induced Stacking Faults in Silicon. I. Nucleation phenomenon. *Journal of Applied Physics* **45**(1), 263–271.

Rozgonyi GA, Deysher RP and Pearce CW 1976 The Identification, Annihilation, and Suppression of Nucleation Sites Responsible for Silicon Epitaxial Stacking Faults. *Journal of The Electrochemical Society* **123**(12), 1910–1915.

Ryuta J, Morita E, Tanaka T and Shimanuki Y 1990 Crystal-Originated Singularities on Si Wafer Surface after SC1 Cleaning. *Japanese Journal of Applied Physics* **29**(10), L1947–L1949.

Schöder DK 1989 Lifetime in Silicon Gettering and defect engineering in the semiconductor technology Sci-Tech pp. 383–394.

SEMI M40 2014 Guide for measurement of roughness of planar surfaces on polished wafers (M40-1114) website. URL http://www.semi.org, https://www.semiviews.org.

SEMI M43 2018 Guide for reporting wafer nanotopography (M43-0418) website. URL http://www.semi.org, https://www.semiviews.org.

SEMI M78 2018 Guide for determining nanotopography of unpatterned silicon wafers for the 130 nm to 22 nm generations in high volume manufacturing (M78-0618) website. URL http://www.semi.org, https://www.semiviews.org.

SEMI MF1188 2018 Test method for interstitial oxygen content of silicon by infrared absorption with short baseline (MF1188-1107) website. URL http://www.semi.org, https://www.semiviews.org.

SEMI MF1239 2016 Test method for oxygen precipitation characteristics of silicon wafers by measurement of interstitial oxygen reduction (MF1239-0305) website. URL http://www.semi.org, https://www.semiviews.org.

SEMI MF1391 2012 Test method for substitutional atomic carmon content of silicon by infrared absorption (MF1391-1107) website. URL http://www.semi.org, https://www.semiviews.org.

SEMI MF1630 2018 Test method for low temperature ft-ir analysis of single crystal silicon for III-V impurities (MF1630-1107) website. URL http://www.semi.org, https://www.semiviews.org.

SEMI MF1809 2010 Guide for selection and use of etching solutions to delineate structural defects in silicon (MF1809-1110) website. URL http://www.semi.org, https://www.semiviews.org.

Shimura F 1994 *Oxygen in Silicon*. Academic Press, London, UK.

Sumino K, Harada H and Yonenaga I 1980 The Origin of the Difference in the Mechanical Strengths of Czochralski-Grown Silicon and Float-Zone-Grown Silicon. *Japanese Journal of Applied Physics* **19**(1), L49–L52.

Theocaris P 1964a Isopachic patterns by the moiré method. *Experimental Mechanics* **4**, 153–159.

Theocaris PS 1964b Moiré patterns of isopachics. *Journal of Scientific Instrument* **41**, 133–138.

Tollenaar D 1945 Moiré: Interferentieverschijnselen bij rasterdruk. *Amsterdam Instituut voor Grafische Techniek*.

Tu J 1988 The diffraction near fields and Lau effect of a square-wave modulated phase grating. *Journal of Modern Optics* **35**(8), 1397–1408.

Válek L and Šik J 2012 Defect Engineering During Czochralski Crystal Growth and Silicon Wafer Manufacturing Modern Aspects of Bulk Crystal and Thin Film Preparation InTech (www.intechopen.com) chapter 3, pp. 43–70.

Wei S and Kao I 1999 Hight-resolution wafer surface topology measurement using phase-shifting shadow moiré technique In *the Proceedings of IMECE'99: DE-Vol 104, Electronics Manufacturing Issues* (ed. Sahay C, Sammakia B, Kao I and Baldwin D), pp. 15–20. ASME Press, Three Park Ave., New York, NY 10016.

Wei S, Wu S, Kao I and Chiang F 1998 Measurement of wafer surface using shadow moiré technique with Talbot effect. *Journal of Electronic Packaging* **120(2)**, 166–170.

Wu S, Wei S, Kao I and Chiang FP 1997 Wafer surface measurements using shadow moiré with Talbot effect *Proceedings of ASME IMECE'97*, pp. 369–376. ASME Press, Dallas, Texas.

Yamagishi H, Fusegawa I, Fujimaki N and Katayama M 1992 Recognition of D Defects in Silicon Single Crystals by Preferential Etching and Effect on Gate Oxide Integrity. *Semiconductor Science and Technology* **7**(1A), A135–A140.

Yoshizawa T and Tomisawa T 1993 Shadow moiré topography by means of the phase-shift method. *Opt. Eng.* pp. 1668–1674.

Young HT, Liao H and Huang H 2006 Surface integrity of silicon wafers in ultra precision machining. *International Journal of Advanced Manufacturing Technology* **29**(3), 372–378.

11

Conclusion

Wafer manufacturing has been a field in which experience and engineering know-how have played significant roles and have provided instrumental value to the industry. With the emergence of new technologies and research, the challenges are to understand the fundamental process modeling in various manufacturing processes of wafer production, and to be able to apply this new knowledge in the process control and management of wafering processes. The underlying principle of this book is to provide research experience and expertise in wafer manufacturing to provide a resource to researchers and practitioners alike in wafer manufacturing.

This book is organized into three categories, with each category including topics that are related to the broader subject of the category. The three categories are:

(I) From crystal to prime wafers
(II) Wafer forming
(III) Wafer surface preparation and management.

Each category includes various topics presented by the chapters and sections that expound the broader subject of the category. They are summarized in the following.

11.1 (I) From Crystal to Prime Wafers

The first category presents the fundamental subject of wafer manufacturing from crystal to prime (or premium) wafers. This is best illustrated in Figure 2.1 in which the generalized process flow of wafer manufacturing from crystal growth to prime wafers is illustrated. Various topics are presented in order. This category I includes Chapters 1 to 3.

First of all, Chapter 1 starts out with a general introduction of wafers and semiconductors, including the semiconductor revolution since the invention of the first transistor at the Bell Lab in 1947; wafers used in device manufacturing (IC and MEMS); topics in surface properties and quality measurements of wafers; other properties of wafers; and the economics of wafer manufacturing. The presentation on the topic of wafer surface specifications encompasses a variety of important wafer surface properties and can serve as an excellent resource. The anisotropy of silicon viewed from different crystalline orientations is discussed with graphical illustrations.

Wafer Manufacturing: Shaping of Single Crystal Silicon Wafers,
First Edition. Imin Kao and Chunhui Chung.
© 2021 John Wiley & Sons Ltd. Published 2021 by John Wiley & Sons Ltd.

Next, the generalized process flow of wafer manufacturing is summarized and illustrated in Figure 2.1, which serves as a reference throughout this book, and should be a ubiquitous reference for wafer manufacturing. The topics in Chapter 2 start with crystal growth processes which introduces various methods of crystal growth of single crystalline (or monocrystalline) and polycrystalline crystals. There are many handbooks and references in the topic of crystal growth. The readers may want to check out the references at the end of this chapter to further explore this topic. Various wafer forming processes are discussed, with comparison among processes serving similar functionality. This is followed by a presentation on wafer surface processing and preparing. Industrial processes of wafer manufacturing are presented with relevant photos of equipment and processes, provided by the Global Wafer Co., Ltd in Taiwan. This is a unique section that puts relevant industrial practice in perspective with the book.

The first category of subject ends with Chapter 3, which expounds on several important and fundamental manufacturing concepts and principles in brittle machining of semiconductor materials. The fundamental differences between ductile and brittle machining are presented. Bonded abrasive machining (BAM) and free abrasive machining (FAM) are explained with their different machine process modeling and effects on the machining of materials. Various standards of abrasive grits are presented and compared. The focus on abrasive machining and process modeling are presented to facilitate the following discussion on individual wafer forming, wafer surface preparation and machining processes.

11.2 (II) Wafer Forming

The second category presents the the technology of forming sliced wafer into the shape necessary for semiconductor fabrication using planar technology. This category II includes Chapters 4 to 6.

Wafer forming is the first post-growth process in wafer manufacturing. Chapter 4 starts the subject of wafer forming by introducing wafer slicing using a modern slurry wiresaw. Since the dawn of the 20th century, the wire saw had been used for stone quarrying. Of course, the modern wire saws for wafer production are subject to a lot more stringent requirements to produce prime wafers of high surface quality. Modern slurry wiresaws were first employed in the US for slicing photovoltaic (PV) wafers in the early 1990s, while slowly being adopted by the semiconductor industry to produce single crystalline silicon wafers. The modern slurry wiresaw is compared with the ID saw, which was replaced by wiresaws. Technology and research issues in wiresaw manufacturing process are presented.

This is followed by Chapter 5 which presents a comprehensive range of contents in the modeling of wiresaws process and material characteristics. The rolling-indenting model of the FAM process of slurry wiresaws is presented, followed by the vibration analysis of a moving wire and its role in the wiresawing process. The practical damping factor is derived to model the damping environment of the wiresaw system. The damping factors are important in accurate modeling for a moving wire in an environment of continuum, described by a partial differential equation. The closed-form solutions of undamped vibration of a moving wire was solved in early 1990; however, the closed-form solution of the damped vibration of a moving wire was not solved until the 2010s. The solution of damped vibration of a moving wire is important and relevant in understanding the slicing operation using a slurry wire-saw. Elasto-hydrodynamic process modeling is another important research to understand the intertwined relationship between the hydrodynamic effect of a fast moving wire in a damped environment with abrasive slurry and the elastic characteristics of the wire. A few degrees of difference in traditional manufacturing processes may not be critical; however, such a difference can be very critical in precision manufacturing in which the change of a few tens of microns can be critical in the control of the outcomes of the process. Therefore, thermal management is important in the wafering process. Finally, the application of moiré optical metrology is presented to monitor real-time wire wear and manage the wire web of wiresaws.

The second category of subject ends with Chapter 6 which presents an emerging and important technology of diamond-impregated wire saws, or diamond saws. Diamond wire saws have been replacing slurry wiresaws in recent years and have become increasing important in slicing materials with higher hardness, such as aluminosilicate glass (used for screen surfaces of mobile phones), sapphire, silicon carbide, and others, in addition to silicon. It is a growing field that still requires more research to understand and improve the slicing processes. The processes to produce diamond-impregated wires, which are critical in the application of diamond wire saws, are presented, along with slicing performance under different process parameters. Properties of wafers produced by diamond saws are discussed, with comparison to the wafers produced by the slurry wiresaws. The diamond wire saw is a topic that has enjoyed growing popularity in usage and is expected to continue to develop with the improvements in the technology of the production of diamond-impregnated wires.

11.3   (III) Wafer Surface Preparation and Management

The third category deals with the various technologies of preparing the wafer surface for premium wafers ready for microelectronics fabrication. This category III includes Chapters 7 to 10.

The third and last category of subject in wafer manufacturing starts with the lapping process in Chapter 7. The fundamentals of FAM process in lapping are presented first, followed by discussions on the single-sided, double-sided and soft-pad lapping processes. The lapping operation is presented with an illustration of lapping equipment and mechanisms for better understanding of the process. Lapping and preliminary planarization of wafer surface are presented next with technical challenges in lapping.

Chapter 8 presents the topic of the chemical mechanical polishing (CMP) process. A schematic of CMP equipment is used to illustrate the different components of an industrial CMP machine and to explain the CMP process. The specifications of polished silicon wafers are presented, in particular the requirements of the flatness of the wafer surface. Polishing pad technology and polishing slurry technology are discussed next. After that, the topic of edge polishing is discussed. Edge polishing can improve the yield in IC fabrication by avoiding the defects along the wafer edge profile to affect the defects of fabrication.

Next, a few topics are collected and presented in Chapter 9, which includes topics in surface grinding, edge grinding, etching and surface cleaning. Surface grinding is an emerging technology with interesting research and technology development. This topic could have been combined with lapping in Chapter 7 to start the category of wafer surface preparation. Edge grinding is a necessary step in early procedure of wafer forming to remove chips and cracks along the peripheral edge of a semiconductor wafer. This process is also called edge rounding in other literature. Topics in etching are presented next, especially the preferential etching and the different etchants developed and standardized in the characterization of defects over the years.

The prime wafers must be prepared with surface cleaning using the famous RCA cleaning process before they go through a final inspection and packaging to ship as premium wafers for microelectronic fabrication. The standard RCA cleaning protocol is presented with techniques and variation of the RCA method. References are provided for further study on this topic.

The third and last category of subject ends with Chapter 10 which introduces innovative optical metrology techniques for a real-time and whole-wafer measurements, called the shadow moiré. Phase shifting technique is presented to augment the resolution of the technique. In addition, wafer surface specifications are outlined with reference to Chapter 1. Various wafer defects, including line defects, planar defect, bulk defects, and others, are presented with their impact on device yield, as well as various inspection systems for detection.

11.4 Final Remarks

The book is a must-have for engineers and researchers in the field of wafer production and manufacturing. Although single crystalline silicon wafers are the majority of wafering done today in industry, this book is also equally applicable, in many regards, to the wafer manufacturing of other crystal ingots such as III–V and II–VI compounds, silicon carbide, lithium niobates, aluminosilicate glass, sapphire, and others.

It has been a major undertaking to commit to writing a book like this, although I have been conducting research in wafer manufacturing since 1994. This project has taken more time than I thought it would when I first signed the agreement with Wiley. My goal was and still is to make this book a resource for those who want to understand the many aspects of wafer manufacturing technologies, as well as for those practitioners who want to utilize the knowledge to build engineering intuition and improve existing processes.

Index

Wafer Manufacturing: Shaping of Single Crystal Silicon Wafers,
First Edition. Imin Kao and Chunhui Chung.
© 2021 John Wiley & Sons Ltd. Published 2021 by John Wiley & Sons Ltd.